AF333764

NEW ASPECTS OF HUMAN POLYMORPHONUCLEAR LEUKOCYTES

Recent Volumes in this Series

NEW ASPECTS OF HUMAN POLYMORPHONUCLEAR LEUKOCYTES

Edited by

W. H. Hörl

Universitätskliniken des Saarlandes
Homburg/Saar, Germany

and

P. J. Schollmeyer

Medizinische Universitätsklinik
Freiburg, Germany

PLENUM PRESS • NEW YORK AND LONDON

Library of Congress Cataloging in Publication Data

New aspects of human polymorphonuclear leukocytes / edited by W. H. Hörl and
P. J. Schollmeyer.
    p.   cm. — (Advances in experimental medicine and biology; v. 297)
    Proceedings of the International Symposium "New Aspects of Human
Polymorphonuclear Leukocytes," held in Freiburg (FRG) on Oct. 26–28, 1989.
    Includes bibliographical references and index.
    ISBN 0-306-43906-9
    1. Neutrophils — Congresses. I. Hörl, Walter H. II. Schollmeyer, P. J. III. Interna-
tional Symposium "New Aspects of Human Polymorphonuclear Leukocytes". (1989:
Freiburg im Breisgau, Germany) IV. Series.
    [DNLM: 1. Neutrophils — congresses. W1 AD559 v. 297 / WH 200 N5315 1989]
QR185.8.N47N52   1991
612.1'12 — dc20
DNLM/DLC   91-3492
for Library of Congress   CIP

Proceedings of a workshop on New Aspects of Human Polymorphonuclear
Leukocytes, held October 26–28, 1989, in Freiburg, Germany

ISBN 0-306-43906-9

© 1991 Plenum Press, New York
A Division of Plenum Publishing Corporation
233 Spring Street, New York, N.Y. 10013

Printed in the United States of America

PREFACE

We are pleased to present our readers the proceedings of the
International Symposium on "New Aspects of Human Polymorpho-
nuclear Leukocytes" which was held in Freiburg im Breisgau,
FRG, from October 26-28th, 1989.

The meeting provided an unique framework for close interaction
between scientists from various disciplines, including biochemistry, biology, physiology, pathology, clinical chemistry,
hematology, gynecology, surgery, intensive care medicine,
nephrology, rheumatology, and infectious diseases.

We would like to express our gratitude and appreciation for
all those who have stimulated, encouraged, and supported us to
hold the symposium in Freiburg. This endeavor could not have
been possible without the generous financial support of Asid-
Bonz (Böblingen), Bayer AG (Leverkusen), Bayropharm GmbH
(Köln), Baxter (München), Ciba-Geigy (Wehr/Baden), Cilag GmbH
(Sulzbach), Fresenius AG (Oberursel), Gambro (Martinsried),
Gry-Pharma GmbH (Kirchzarten), Hoechst AG (Frankfurt), Hospal
(Nürnberg), Knoll AG (Ludwigshafen), Lederle-Cyanamid
(Wolfratshausen), E. Merck (Darmstadt), MSD Sharp and Dohme
GmbH (München), Pfizer GmbH (Karlsruhe), and Kabi/Pfrimmer
(Erlangen).

We are indebted to Mrs. I. Szkibik for her invaluable assistance both with the organization of the meeting and the preparation of the manuscripts.

Walter H. Hörl
Peter J. Schollmeyer

CONTENTS

ELECTROPHYSIOLOGICAL PROPERTIES OF HUMAN NEUTROPHILS

Karl-Heinz Krause[+], Daniel P. Lew[+], Micheal J. Welsh[*]

[+]Division of Infectious Diseases,
University Hospital, 1211 Geneva 4, Switzerland
and [*]Howard Hughes Medical Institute, University
of Iowa College of Medicine, Iowa City, Iowa
52242, U.S.A.

INTRODUCTION

Plasma membrane ion channels play an important role in
the regulation of cellular activity and as effector proteins
for many cellular functions. Despite substantial progress in
the understanding of signal transduction and cellular
activation in neutrophils, little is known about the nature
and properties of ion channels in these cells. The intracellular ion concentrations of a neutrophil under physiological conditions are shown in Figure 1.

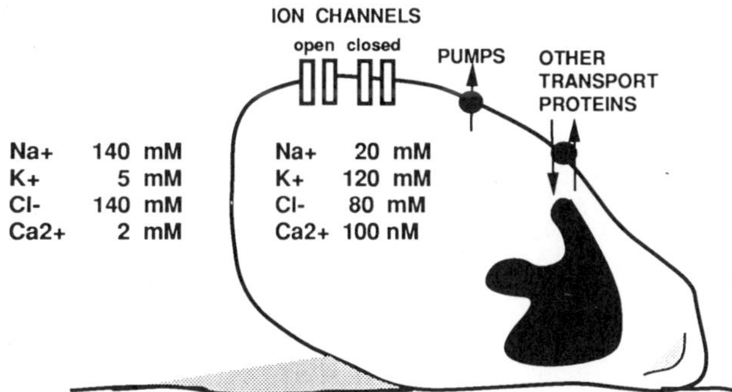

Figure 1. <u>Intra- and extracellular ion concentrations under
physiological conditions for a neutrophil.</u> Ionic gradients,
transmembrane potential, and ion fluxes during cellular
activation are determined by channels, pumps and other
transport proteins (e.g. cotransporters, exchangers, etc.)
in the plasma membrane.

New Aspects of Human Polymorphonuclear Leukocytes
Edited by W.H. Hörl and P.J. Schollmeyer, Plenum Press, New York, 1991

As observed in most other cell types, the intracellular
concentrations of various ions are different from those
observed in the extracellular medium. Intracellular ion
concentrations are high for K^+, low for Na^+ and extremely
low for Ca^{2+}. The intracellular Cl^- concentration in human
neutrophils is around 80 mM (Simchowitz 1986), thus lower
than extracel-lular Cl^- (140 mM), but relatively high
compared to other cells (Hille 1984). Ionic gradients, like
the one observed in neutrophils, are determined by the
transport proteins in the cell membrane, including pumps,
transporters and channels. Pumps use cellular sources of
energy (e.g. ATP) to transport ions against electrochemical
gradients. Other membrane transporters perform ion exchange,
cotransport, and countertransport to move ions across
membranes. Their capacity is relatively low (maximum 103
ions/sec). Channels, in contrast, are pores that allow
passive movement of ions down favorable electrochemical
gradients. They can be in an open (=active) or closed state.
Channels are often ion selective and thus provide a mean of
making the plasma membrane conductive for specific ions.
When compared to other transport proteins, ion channels have
a very high capacity (at least 106 ions/sec).

Channels have a variety of functions in different cell
types (for review see Hille 1984). Many cells have K+
channels that are active in the resting cell and lead to K^+
efflux because of the gradient between the high intracellular and the low extracellular K^+ concentration. If the
cell membrane is relatively impermeant for other ions, K^+
efflux can not be counterbalanced and leads to build-up of a
negative charge in the cell interior, i.e. a negative
resting potential. In so-called excitable tissues (nerve,
muscle, etc.), stimulation of cells opens Na^+ channels; Na^+
influx depolarizes the cell and leads to an action potential. A variety of cells have different types of Ca^{2+}
channels that allow Ca^{2+} influx and an increase in the
cytosolic free Ca^{2+} concentration during cellular activation. Many cells have Cl^- channels which may be involved in
regulation of membrane potential (e.g. GABA-receptor in
brain, Hille 1984) fluid secretion (epithelial cells, Welsh
1988) or cellular volume homeostasis (Hoffmann 1978,
Grinstein et al. 1982, Christensen 1987, Yamaguchi 1989).

Plasma membrane channels of small cells, such as
neutrophils, can be studied directly using the patch clamp
technique or by indirect methods. The indirect methods
include radioactive tracer fluxes, determination of intracellular ion concentrations with fluorescent dyes or measurement of plasma membrane potential with charged lipophilic compounds. The three indirect methods, however, give
only limited information about the ion channels of a cell.
Using radioactive tracer fluxes it is difficult to
distinguish between transport processes and conductive
pathways. Changes of intracellular ion concentrations are
the net result of the activity of all the different transport and conductive pathways for a certain ion. Plasma
membrane potential reflects the sum of all electrogenic
processes across the plasma membrane. The patch clamp
method, in contrast, allows direct determination of plasma
membrane ionic conductances and the study of their
regulation.

In this article we will review recent data on the electrophysiological properties of human neutrophils, in particular:

1. Studies on membrane potential in neutrophils;
2. Whole cell patch clamp studies in neutrophils;
3. Studies on Ca^{2+} influx in neutrophils;
4. Comparison of membrane potential and ion channels in macrophages and neutrophils.

1. STUDIES ON MEMBRANE POTENTIAL IN NEUTROPHILS

Using membrane potential sensitive fluorescent dyes or radiotracers the membrane potential in populations of neutrophils has been studied. Efforts have been made to determine the resting potential, the potential changes during cellular activation, and the ionic basis of those phenomena.

The resting potential of neutrophils has been estimated between -20 and -105 mV (Korchak and Weissman 1978, Kuroki et al. 1982, Seligman et al. 1980, Bashford and Pasternak 1985, Martin, Nauseef, and Clark 1988, Henderson, Chappel and Jones, 1987, Simchowitz and De Weer 1986). These large deviations show that the method is poorly suited for the determination of absolute values. Still, most groups now agree on resting values between -50 and -60 mV. An increase in the extracellular K^+ concentration depolarizes neutrophils (Seligmann et al. 1980). This suggests that, similar to other cell types (see introduction), the resting potential in neutrophils is maintained by a K^+ channel.

Membrane potential changes during cellular activation in neutrophils have been described as follows:

a) Neutrophils depolarize in response to receptor agonists (Seligman et al. 1980), but also in response to agonists that bypass the receptor mechanism, e.g. ionomycin, a calcium ionophore, or phorbol myristate acetate, a protein kinase C activator (Krause et al. 1985).

b) Membrane depolarization in response to receptor agonists is inhibited by preincubation of cells with pertussis toxin (Krause et al. 1985). This argues against a mechanism of depolarization that is directly linked to the receptor (as it may be in excitable tissues) and suggests mediation by a G-protein.

c) The ionic basis of the depolarizing current in neutrophils has not yet been identified. Replacement of extracellular Na^+ by choline does not affect depolarization (Kuroki et al. 1982). As choline is a cation that does not usually pass through Na^+-selective channels, this observation indicates that Na^+ channels are not required for depolarization in neutrophil. However, in other cell types non-selective channels that conduct choline have been described and a role of non-selective

channels in neutrophil depolarization has not yet been
excluded. In some cell types closing of K^+ channels or
opening of Cl^- channels is thought to depolarize the
cells (Cook 1988, Welsh 1988). These possibilities have
not yet been studied in neutrophils. It has also been
proposed that the electron-transfer from the cytosol to
the extracellular space that occurs during the respiratory burst of neutrophils is electrogenic and leads to
plasma membrane depolarization (Henderson, Chappell, and
Jones 1987).

The physiological role of depolarization in neutrophils
is different from excitable cells. Neither Ca^{2+} influx nor
cellular activation is observed after depolarization of
neutrophils by high extracellular K^+ or by Na^+/K^+ ionophores
such as gramicidine (Andersson et al. 1986, Di Virgilio et
al. 1987). In contrast, if depolarization precedes cellular
stimulation by chemoattractants, the increase in cytosolic
Ca^{2+} and cellular responses are diminished. A role of
depolarization as negative feed back during neutrophil
activation has therefore been proposed (Di Virgilio 1987).

2.   WHOLE CELL PATCH CLAMP STUDIES IN HUMAN NEUTROPHILS

The patch clamp was initially designed as a method to
study single ion channels. However, it rapidly emerged as a
technique with a variety of applications, such as the
measurement of whole cell currents in small cells, the study
of interactions of intracellular messengers with ion channels, and the study of control of secretion (Hamill et al.
1981, Neher 1989). The whole cell patch clamp technique measures the channel currents from the entire plasma membrane
of a single cell, whereas the single channel patch clamp
technique measures the current in a randomly chosen small
piece of membrane on the non-adherent side of the cell.
Accordingly, the whole cell patch clamp technique, as compared to single channel measurements, is better suited for the
initial study of cells whose set of ion channels is not yet
known. We therefore used the whole cell patch clamp to study
transmembrane currents in adherent neutrophils. As the cytosolic free Ca^{2+} concentration, $[Ca^{2+}]_i$, is known to be an
important intracellular messenger in neutrophils, we also
studied the modification of whole cell currents by changes
in $[Ca^{2+}]_i$. We found voltage-dependent K^+ channels and
Ca^{2+}-activated K^+ and Cl^- channels. (If not stated other-
wise, the original reference is Krause and Welsh, 1990).
Some properties of these channels are described below.

2.1. VOLTAGE DEPENDENT K^+ CHANNEL

A voltage activated current was found in unstimulated
neutrophils studied with the whole cell patch clamp technique. This channel had a threshold of voltage activation of
-60 mV. Maximal activation was observed at +90 mV. No time-
dependent inactivation was found, even at large depolarizing voltages. Tail current analysis under varying ionic
conditions suggested that the channel was selective for K^+.
The channel current showed inward rectification, i.e. a

larger conductance for K^+ influx than for K^+ efflux. The
channel was blocked by $BaCl_2$, but not by 5 amino-pyridine, a
blocker of the depolarization activated K^+ channel of macrophages.

The properties of the channel suggest that it may
determine the resting membrane potential of neutrophils as
a) it is found in unstimulated cells, b) it is K^+ selective
and thus generates a negative plasma membrane potential by
electrogenic K^+ efflux, and c) it has a threshold of voltage
activation of -60 mV and would therefore hyperpolarize neutrophils only to this voltage, i.e. the membrane potential
of resting neutrophils (a K^+ channel that is independent of
voltage would be expected to hyperpolarize cells to -80 mV,
the reversal potential of K^+).

2.2. Ca^{2+}-ACTIVATED K^+ AND Cl^- CHANNELS IN NEUTROPHILS

Activation of neutrophils by various receptor agonists
leads to an elevation of $[Ca^{2+}]_i$. In other cell types, such
$[Ca^{2+}]_i$ elevations activate a variety of channels. It was
therefore of interest to study the effect of Ca^{2+} on channel
currents in neutrophils. We exposed neutrophils in the whole
cell patch configuration to the Ca^{2+} ionophore ionomycin.
This procedure led to an increase in whole cell currents,
thus demonstrating the presence of Ca^{2+}-activated ion channels in neutrophils. Analysis of the ionic selectivity of
the Ca^{2+}-activated channels revealed the existence of both,
Cl^- and K^+ channels.

As many K^+ channels are voltage dependent and Ca^{2+}
activated, we considered the possibility that the depolarization-activated K^+ channel (described in section 2.1.) and
the Ca^{2+}-activated K^+ conductances reflect two modes of
activation of the same channel. However several observations argue against this possibility. The voltage activated
channel appeared to be independent of intracellular Ca^{2+}
concentrations, while the Ca^{2+}-activated K^+ channel appeared
to be independent of voltage. In addition, the two K^+ channels could be distinguished by their conductive properties.
The voltage activated K^+ channel was inwardly rectifying,
i.e. K^+ ions passed more readily from the outside of the
cell to the inside than from the inside to the outside. The
Ca^{2+}-activated K^+ channel, in contrast, was outwardly rectifying.

The effect of a concomitant Ca^{2+}-induced opening of Cl^-
and K^+ channels on net salt fluxes in a neutrophil would be
a loss of KCl to the extracellular space. Such a loss of
intracellular KCl has been shown in other cell types to be
accompanied by a loss of intracellular water and a decrease
in cell volume (Hoffmann 1978, Grinstein et al. 1982,
Christensen 1987, Yamaguchi 1989).

What could be the physiological significance of a Ca^{2+}-
induced decrease of cell volume? In other cell types a
Ca^{2+}-conductive pathway sensitive to cell volume has been
described (Christensen 1987, Yamaguchi 1989). Exposure of
such cells to hypotonic solutions leads to the following

cascade of events: increase in cell volume; activation of volume sensitive Ca^{2+} influx; activation of Ca^{2+}-gated K^+ and Cl^- channels; net loss of KCl and water; decrease of cell volume towards normal. Neutrophils might have such a mechanism, as they are resistant to hypotonic solutions: brief exposure to distilled water, a routine step in neutrophil purification which lyses red blood cells, does not damage neutrophils. However, no volume-sensitive Ca^{2+} influx has yet been described in neutrophils.

Alternatively, changes in cell volume might be a part of the response to the increase in intracellular Ca^{2+} that occurs during cellular activation. Many neutrophil functions that are mediated or accompanied by rises in intracellular Ca^{2+}, such as chemotaxis, adherence to surfaces and spreading, pseudopode formation and phagocytosis, might necessitate changes in cell volume. The possible involvement of Ca^{2+}-activated Cl^- and K^+ channels in these neutrophil functions will be an important subject of further studies.

3. STIMULATED Ca^{2+} INFLUX IN HUMAN NEUTROPHILS

A rise of the cytosolic free Ca^{2+} concentration is an important intracellular messenger in neutrophils (Pozzan et al. 1983). It is thought to consist of two components, a release of Ca^{2+} from intracellular stores and an influx of Ca^{2+} across the plasma membrane.

The existence of Ca^{2+} influx in neutrophils is suggested by the observations that chemoattractant induced increases in Ca^{2+} are attenuated and markedly shortened in the absence of extracellular Ca^{2+} (Pozzan et al. 1983) and that Mn^{2+}, a divalent cation that can enter cells through Ca^{2+}-selective pathways, enters neutrophils after stimulation by chemotactic peptides (Andersson et al. 1986).

The following properties of Ca^{2+} influx in neutrophils have been described:

a) As opposed to excitable tissues, Ca^{2+} influx in neutrophils is not triggered by plasma membrane depolarization and is not inhibited by blockers of voltage dependent Ca^{2+} channels, such as nifedipine (Andersson et al. 1986). Thus, neutrophils do not possess voltage dependent Ca^{2+} channels;

b) All cell surface receptor agonists, described so far, that induce phosphatidyl-inositol turnover, generation of inositol 1,4,5-trisphosphate and release of Ca^{2+} from intracellular stores also induce Ca^{2+} influx in neutrophils;

c) Chemotactic peptide induced Ca^{2+} influx is inhibited by pertussis toxin to a similar extend as Ca^{2+} release from intracellular stores (Krause et al. 1985). As pertussis toxin is thought to inhibit signal transduction by certain G-proteins, this finding argues against a role of the chemotactic peptide receptor as ion channel. It suggests that the mechanism of Ca^{2+} influx is coupled to the receptor by a G-protein, directly or via a second messenger;

d)   Experiments comparing the time course of Ca^{2+} influx and
     inositol phosphate production showed good correlation
     between elevations of inositol 1,3,4,5 tetrakisphosphate
     and Ca^{2+} influx, suggesting a link between these two
     events (Pittet et al. 1989).

     Data from other cell types suggest the involvement of
inositol phosphates as second messengers that mediate Ca^{2+}
influx. In mast cells (Penner, Matthews and Neher 1988),
lacrimal gland cells (Morris et al. 1987, Llano, Mary, and
Tanguy 1987) and Xenopus leavis oocytes (Snyder, Krause and
Welsh 1988), application of inositol 1,4,5-trisphosphate by
whole cell patch clamp or by microinjection leads to Ca^{2+}
influx. One group reported the need for concomitant inositol
1,3,4,5-tetrakisphosphate injection for the induction of
Ca^{2+} influx in lacrimal cells (Morris et al. 1987). It has
also been proposed that inositol 1,4,5-trisphosphate might
exclusively act by emptying the intracellular Ca^{2+} pool and
that the filling state of the intracellular Ca^{2+} pool regu-
lates Ca^{2+} influx (Takemura and Putney 1989).

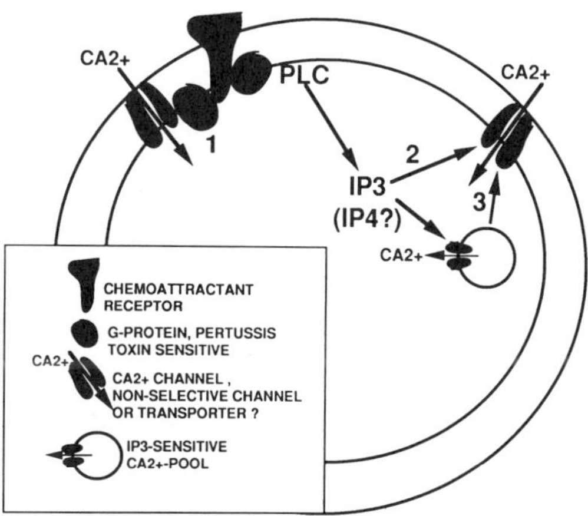

Figure 2. <u>Possible pathways of receptor operated Ca2+ influx
in neutrophils.</u> Ca^{2+} might enter neutrophils through
Ca^{2+}-selective channels, non-selective channels or a
transport pathway. The mechanism of Ca^{2+} influx might be
coupled to the chemoattractant receptor directly by a G
protein (pathway 1). Ca^{2+} influx might also be coupled to
the receptor via phospholipase C (PLC) activation, inositol
1,4,5 trisphosphate (IP3) and eventually inositol 1,3,4,5
tetrakisphosphate (IP4) production (pathway 2).
Alternatively it has been proposed that the emptying of
intracellular Ca^{2+} stores by IP3 is the signal for Ca^{2+}
influx (pathway 3).

Virtually nothing is known about the mechanism of Ca^{2+} influx in neutrophils and the conductive nature of the pathway has not convincingly been demonstrated. An Ins 1,4,5 P3-sensitive Ca^{2+}-conducting plasma membrane channel has been described in T-lymphocytes (Kuno and Gardner 1987), however the activity state of a similar channel in mast cells did not correlate with Ins 1,4,5-P3 induced Ca^{2+} influx as measured by fluorescent dyes (Penner, Matthews and Neher 1988). Whole cell patch clamp studies in mast cells could not reveal a significant Ca^{2+} current (Penner, Matthews and Neher 1988). In single channel patch clamp studies in neutrophils a Ca^{2+}-activated non-selective cation channel has been described (Von Tscharrner et al. 1986). This channel was able to conduct Ca^{2+}, however its role as the physiologically important Ca^{2+} channels of neutrophils is questionable as Ca^{2+} influx and Ca^{2+} release from internal stores can be temporally separated (Irvine 1987). Thus the question, of whether Ca^{2+} enters neutrophils (or related cells) through a Ca^{2+}-selective channel, a non-selective channel or some non-conductive transport pathway remains open (Fig. 2).

4. COMPARISON OF MEMBRANE POTENTIAL AND ION CHANNELS IN MACROPHAGES AND NEUTROPHILS

Macrophages are cells that share a variety of properties with neutrophils. It is therefore of interest to compare the results of electrophysiological studies in neutrophils with those obtained in macrophages (Fig. 3, Gallin 1988, Krause and Welsh 1990). Resting membrane potential of macrophages and neutrophils has been reported to be in a similar range (-50 to -60 mV). Voltage dependent K^+ channels in both cell types, macrophages (Gallin 1988) and neutrophils (Krause and Welsh 1990) are likely to maintain this resting potential. However, the neutrophil channel shows distinct properties. It differs from the macrophage channel because: it does not deactivate with time, it is not inhibited by 4 amino-pyridine and it shows inward rectification.

After stimulation with chemotactic peptides or Ca^{2+} ionophores, neutrophils depolarize, while macrophages hyperpolarize. Two types of Ca^{2+}-activated K^+ channels have been described in macrophages, a voltage dependent type and a voltage-independent type. A $[Ca^{2+}]_i$ increase in macrophages should activate these two channels and thereby lead to hyperpolarization. In addition macrophages possess a hyperpolarization activated K^+ channel which might provide a further positive feed back to the Ca^{2+}-induced hyperpolarization. Neutrophils have only one kind of Ca^{2+}-activated K^+ channel and no hyperpolarization activated K^+ channel. However, they have a Ca^{2+}-activated Cl^- channel that has not been described in macrophages. Could this Ca^{2+}-activated Cl^- channel play a role in the depolarization produced during neutrophil activation? The intracellular Cl^- concentration in neutrophils is relatively high (80 mM, Simchowitz and de Weer 1986). The reversal potential for Cl^- is therefore less negative than the resting plasma membrane potential (-14 mV versus -60 mV). Activation of Cl^- channels by Ca^{2+} could therefore depolarize neutrophils, depending on the relative contribution of Ca^{2+}-activated Cl^- channels and Ca^{2+}-activated K^+ channels. However depolarization of neutrophils in response to physiological stimuli is only partially Ca^{2+}-dependent and Ca^{2+}-independent depolarization is observed in response to phorbol esters. Thus, while Ca^{2+} activation of Cl^- channels might contribute to depolarization in neutrophils, it is unlikely to be the only mechanism.

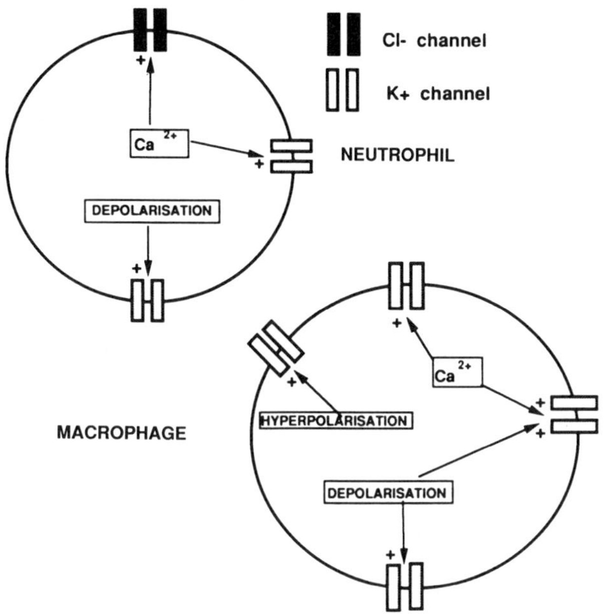

Figure 3. <u>Comparison of ion channels in neutrophils and macrophages.</u> Neutrophils have voltage dependent K^+ channels and Ca^{2+} activated K^+ and Cl^- channels. Macrophages have 4 types of K^+ channels: a depolarization-activated channel, a hyperpolarization-activated channel, a Ca^{2+}- and depolarization-activated channel and a Ca^{2+}-activated, voltage-independent channel.

CONCLUSION

Neutrophils have a clearly distinct set of ionic channels, even when compared to those found in related cells like macrophages. The analysis of the role of these channels in the regulation of cellular function will contribute to our understanding of the specific features of stimulus response coupling in neutrophils. Ion channels might also be used as targets for cell specific pharmacological interventions.

Acknowledgements: This work was supported by a grant of the Swiss National Foundation 3.829.0.87 and the NIH (HL 29851).

REFERENCES

1. Andersson, T., Dahlgren, C., Pozzan, T., Stendahl, O., and Lew, P.D.: Characterization of fmet-leu-phe receptor-mediated Ca^{2+} influx across the plasma membrane of human neutrophils. _Mol. Pharmacol._ 30:437 (1986).

2. Bashford, C.L., Pasternak, C.A.: Plasma membrane potential of neutrophils generated by the Na^+ pump. _Biochim. Biophys. Acta_ 817:174 (1985).

3. Christensen, O.: Mediation of cell volume regulation by Ca^{2+} influx through stretch-activated channels. _Nature_ 330:66 (1987).

4. Cook, N.S.: The pharmacology of potassium channels and their therapeutic potential. _Trends in Pharmacology_ 9:21 (1988).

5. Grinstein, C.A., Clarke, C.A., Dupre, A., Rothstein, A.: Volume induced increase of anion permeability in human lymphocytes. _J. Gen. Physiol._ 80: 801.

6. Hamill, O.P., Marty, A., Neher, E., Sakman, B., Sigworth, F.: Improved patch-clamp techniques for high-resolution current recording from cells and cell-free membrane patches. _Pflugers Arch._ 391:85 (1981).

7. Henderson, L.M., Chappell, J.B., Jones, O.T.G.: The superoxide-generating NADPH oxidase of human neutrophils is electrogenic and associated with an H^+ channel. _Biochem. J._ 246:325 (1987).

8. Hille, B.: Ion channels in excitable membranes. Sinauer Associates Inc. Sunderland, Massachusetts. p. 426 (1984).

9. Hoffman, E.: Regulation of cell volume by selective changes in the permeabilities of Ehrlich ascites tumor cells. In: Osmotic and Volume Regulation. Alfred Benzon Symposium XI. C.B. Jorgensen and E. Skadhauge, eds. p. 397 (1978).

10. Irvine, R.F.: Inositol phosphates and calcium entry. _Nature_ 328:386 (1987).

11. Korchak, H.M., Weissmann, G.: Changes in membrane potential of human granulocytes antecede the metabolic responses to surface stimulation. _Proc. Natl. Acad. Sci. USA_ 75: 3818-3822 (1978).

12. Krause, K.H., Schlegel, W., Wollheim, C.B., Andersson, T., Waldvogel, F.A., Lew, P.D.: Chemotactic peptide activation of human neutrophils and HL-60 cells. Pertussis toxin reveals correlation between inositol trisphosphate generation, calcium ion transients and cellular activation. _J. Clin. Invest._ 76:1348 (1985).

13. Krause, K.H., Welsh, M.J.: Voltage dependent and Ca^{2+}-activated ion channels in human neutrophils. _J. Clin. Invest._ in press (1990).

14. Kuno, M., Gardner, P.: Ion channels activated by inositol 1,4,5-triphosphate in plasma membrane of human T-lymphocytes. _Nature_ 326:301 (1987).

15. Kuroki, M., Kamo, N., Kobatake Y., Okimasu, E., Utsumi,

K.: Measurement of membrane potential in polymorpho-nuclear leukocytes and its changes during surface stimulation. <u>Biochim. Biophys. Acta</u> 693:326 (1982).

16. Llano, I., Marty, A., Tanguy, J.: Dependence of intracellular effects of GTP-gamma-S and inositol-trisphosphate on cell membrane potential and on external Ca^{2+} ions. <u>Pflugers Arch.</u> 409:499 (1987).

17. Martin, M.A., Nauseef, W.M., Clark, R.A.: Depolarization blunts the oxidative burst of human neutrophils. Parallel effects of monoclonal antibodies depolarizing buffers, and glycolytic inhibitors. <u>J. Immunol.</u> 140:3928 (1988).

18. Morris, A.P., Gallacher, D.V., Irvine, R.F., Petersen, O.H.: Synergism of inositol trisphosphate and tetra-kisphosphate in activating Ca^{2+}-dependent K^+ channel. <u>Nature</u> 330:653 (1987).

19. Nasmith, P.E., Grinstein, S.: Are Ca^{2+} channels in neutrophils activated by a rise in cytosolic free Ca^{2+}? <u>FEBS Lett.</u> 221:95 (1987).

20. Neher, E.: The use of the patch clamp technique to study second messenger-mediated cellular events. <u>Neuroscience</u> 26:727 (1989).

21. Penner, R., Matthew, G., Neher, E.: Regulation of calcium influx by second messengers in rat mast cells. <u>Nature</u> 334:499 (1988).

22. Pittet, D., Lew, P.D., Mayr, G.W., Monod, A., Schlegel, W.: Chemoattractant receptor promotion of Ca^{2+} influx across the plasma membrane. A role for cytosolic free calcium elevations and inositol 1,3,4,5-tetrakis-phosphate production. <u>J. Biol. Chem.</u> 264:7251 (1989).

23. Pozzan, T., Lew, P.D., Wollheim, C.B., Tsien, R.: Is cytosolic free Ca^{2+}-concentration regulating neutrophil activation? <u>Science</u> 221:1413 (1983).

24. Snyder, P., Krause, K.H., Welsh, M.J.: Inositol trisphosphate isomers, but not inositol 1,3,4,5-tetra-kisphosphate induce calcium influx in xenopus laevis oocytes. <u>J. Biol. Chem.</u> 263:11048 (1988).

25. Seligman, B.E., Gallin, E.K., Martin, D.L., Shain, W., Gallin, J.I.: Interaction of chemotactic factor with human polymorphonuclear leukocytes: studies using a membrane potential-sensitive cyanine dye. <u>J. Membr. Biol.</u> 53:257 (1980).

26. Simchowitz, L., De Weer, P.: Chloride movements in human neutrophils: diffusion, exchange and active transport. <u>J. Gen. Phys.</u> 88:167 (1986).

27. Takemura, H., Putney, J.W.: Capacitative calcium entry in parotid acinar cells. <u>Biochem. J.</u> 258:409 (1989).

28. Von Tscharrner, V., Prod'hom, B., Baggiolini, M., Reuter, H.: Ca^{2+}-activated ion channels in human neutro-phils. <u>Nature</u> 324:369 (1986).

29. Di Virgilio, F., Lew, P.D., Andersson, T., Pozzan, T.: Plasma membrane potential modulates chemotactic peptide-stimulated cytosolic free Ca^{2+} changes in human neutrophils. <u>J. Biol. Chem.</u> 262:4574 (1987).

30. Welsh, M.J.: Electrolyte transport by airway epithelia. <u>Physiol. Rev.</u> 67:1143 (1987).

31. Yamaguchi, D.T., Gree, J., Kleeman, C.R., Muallem, S.: Characterization of volume-sensitive, calcium permea-ting pathways in the osteosarcoma cell line UR-106-01. <u>J. Biol. Chem.</u> 264:4383 (1989).

MECHANISMS OF NEUTROPHIL AND MACROPHAGE MOTILITY

Francesco Di Virgilio, Paola Pizzo and Enzo
Picello

C.N.R. Center for the Study of the Physiology of
Mitochondria and Institute of General Pathology
Via Trieste 75, I-35131 Padova, Italy

INTRODUCTION

Physiological responses of neutrophils and macrophages
are crucially dependent on their motility as they undergo
random or directed locomotion, pinocytosis, phagocytosis and
exocytosis (granule movement). Each of these functions
requires complex interactions between surface receptors, the
cytoskeleton and the plasma membrane (Silverstein et al.,
1977).

Useful insights on the mechanism of leukocyte motility
are derived from studies performed in other ameboid cells such
as Acanthamoeba and Dictyostelium which are more amenable to
genetic manipulation and therefore easier to study in detail.

MOLECULAR COMPONENTS OF THE CONTRACTILE APPARATUS

In the last two decades considerable evidence has accumulated to implicate a pivotal role for the cortical cytoplasm
in the motile responses of leukocytes. This rim of gelled
cytoplasm underlying the plasma membrane is defined as "organelle exclusion zone" to stress the absence of granules,
mitochondria and other intracellular vesicles (Griffin et al.,
1976). Actin filaments and actin-binding proteins are concentrated in the cortical cytoplasm while other cytoskeletal
proteins are mostly restricted to the cell body (Stossel et
al., 1988). Immunofluorescent or NBD-phallacidin staining of
resting or activated phagocytes has also confirmed the active
role of the actin scaffolding in cell motility. In resting
cells actin filaments are homogeneously diffused throughout
the peripheral cytoplasm, but as soon as the cell engages in a
motile response, such as phagocytosis, actin becomes concentrated below the site of attachment of the particle, forming
the so-called "phagocytic cups" (Wang et al., 1984; Sheterline
et al., 1986; Silverstein et al., 1989). Both leukocyte
motility and assembly of the actin network can be inhibited in
parallel by the fungine metabolites cytochalasins.

New Aspects of Human Polymorphonuclear Leukocytes
Edited by W.H. Hörl and P.J. Schollmeyer, Plenum Press, New York, 1991

Leukocytes contain the basic contractile elements present
in striated and smooth muscle cells. Conventional multimeric
myosin II can be extracted from macrophages and neutrophils
and there is evidence that phosphorylation of the 15 kDa
myosin light chain can be activated by chemotactic stimuli
(Trotter et al., 1985). Given this purported role of conventional myosin in leukocyte motility it was surprising that
disruption of myosin II in Dictyostelium either genetically or
by injection of inhibitory antibodies had little effect on
motility, pseudopod protrusion and phagocytosis (Knecht and
Loomis, 1987; De Lozanne and Spudich, 1987).

By far, however, the most abundant cytoskeletal protein
is actin. This protein is highly conserved and present in the
cytoplasm as a monomeric globular protein (G actin, Mr 42
kDa) or as a polymeric double helical filament (F actin).
Under steady state conditions actin monomers exchange with the
opposite filament ends at different rates, the exchange being
much faster at the "barbed" than at the "pointed" end. The
regulation of the elongation of actin filaments is obviously
crucial for cell motility (Stossel, 1988). A number of regulatory actin-binding proteins have been isolated: profilin,
gelsolin, acumentin, alfa-actinin, a 540 kDa Mr homodimer
actin-binding protein and a 42 kDa Ca^{2+}-binding protein that
binds to the barbed end of actin filaments. Profilin (Mr 20
kDa) binds actin monomers with an affinity that can be modulated by phosphatidylinositol 4,5-bisphosphate (PtdIns4,5P$_2$[1])
and it could be important in sequestering unpolymerized actin
(Lassing and Lindberg, 1985). Gelsolin (Mr 84,000) binds the
barbed end of actin and prevents monomer exchange. Gelsolin
also severs intrafilament actin-actin bonds, thereby fragmen-
ting the filaments. Both the severing and barbed end-blocking
activities of gelsolin are activated by micromolar Ca^{2+}
concentrations. In <u>in vitro</u> studies, however, Ca^{2+}-induced
gelsolin-actin interaction is irreversible (EGTA-resistant).
This feature of gelsolin-actin interaction has been a for-
midable obstacle for the proposed role of gelsolin in the
regulation of actin assembly until Janmey and Stossel (1987)
demonstrated that PtdIns4,5P$_2$ could dissociate gelsolin from
actin also in the presence of EGTA. These authors also demonstrated that PtdIns4,5P$_2$ inhibits actin filament severing by
gelsolin even in the presence of micromolar Ca^{2+}.

ACTIN BINDING TO THE PLASMA MEMBRANE

It is obvious that for the cell to be capable of perfor-
ming any kind of controlled movement, a contractile apparatus,
such as the assembling/disassembling actin scaffolding, is a
necessary but not sufficient requirement. The actin filaments
must be linked somehow to the plasma membrane in order to
cause retraction or extension of pseudopods. The interaction

---

[1]Abbreviations: PtdIns4,5P$_2$, phosphatidyl inositol 4,5-
bisphos-phate; Ins1,4,5P$_3$, inositol 1,4,5-trisphosphate;
$(Ca^{2+})_i$, cyto-solic free Ca^{2+} concentration; DAG, diacylglycerol; PMA, phorbol 12-myristate 13-acetate.

between actin and the plasma membrane is one of the major
unsolved problems of leukocyte motility. Electron microscopy
studies of phagocytes have failed to reveal structures similar
to the actin-spectrin aggregates present on the cytoplasmic
surface of red blood cells. Furthermore it is the barbed end
that faces the membrane, raising the problem of how actin
monomers can exchange at this end of the filament (Stossel,
1988). Likewise the direct association of actin with various
membrane receptors (e. g. complement receptor 1, complement
receptor 3 or Fc receptor of human neutrophils) is still of
dubious relevance. Recently, Wuestenube and Luna (1987) have
isolated a 17 kDa protein from the plasma membrane of Dictyostelium named "ponticulin" that might mediate F-actin binding
to the plasma membrane of these cells. The possibility of a
direct linkage between the cytoskeleton and the plasma
membrane has been strengthened by the discovery and characterization of a second type of myosin (now called myosin I)
and by the demonstration of its direct interaction with the
phospholipids of the plasma membranes of Dictyostelium and
Acanthamoeba (Adams and Pollard, 1989; Fukui et al., 1989).
Myosin I is single-headed and, at variance with myosin II,
lacks the alfa helical tail at the carboxy-terminal. The tail
is replaced with a non-helical actin-binding site which is
sensitive to ATP. In vitro studies by Adams and Pollard and by
Korn's group have recently demonstrated the direct association
of myosin I to amoeba plasma membrane vesicles stripped of all
actin, myosin and other peripheral proteins. Even more interesting is the observation that the association between myosin
I and the plasma membrane may be regulated by PtdIns4,5P$_2$. If
this association is confirmed in the intact cell, an important
link would be made between the powerful signalling system of
the phosphatidyl inositol lipids and the cytoskeleton.

REGULATION OF ACTIN ASSEMBLY

The role of cytoplasmic calcium

By analogy to striated muscle contraction and to an over-
whelming number of receptor-mediated responses in eukariotic
cells, Ca^{2+} has been so far believed to be the link between
occupation of plasma membrane receptors and the actin cytoskeleton. However, the association of motile responses to changes
in (Ca^{2+})$_i$ is still a matter of hot debate (see Di Virgilio et
al., 1990), for recent review). Most soluble chemoattractants
are coupled to increases in (Ca^{2+})$_i$ in both neutrophils and
macrophages. More complex is the situation with particulate
stimuli. It was initially shown (Campbell and Hallet, 1983)
that latex particles were ingested by neutrophils in the
absence of any detectable increase in (Ca^{2+})$_i$, using obelin or
quin2 as indicators; later studies, however, indicated that
ingestion of IgG or C3b/C3bi coated particles by neutrophil or
macrophage suspensions was concomitant with an increase in
(Ca^{2+})$_i$ (Lew et al., 1985; Young et al., 1985; Di Virgilio et
al., 1988a). Micro-fluorimetric studies of quin2 for fura-2-
loaded neutrophils and eosinophils also revealed heterogeneity
of (Ca^{2+})$_i$ distribution as a consequence of stimulation with
chemoattractants (Sawyer et al., 1985, Brundage and Fay,
1989). On the contrary inflammatory macrophages plated on
glass coverslips showed no increase in (Ca^{2+})$_i$ when challenged

with IgG-coated erythrocytes (McNeil et al., 1986; Di Virgilio et al., 1988a). Association of motile responses to changes in $(Ca^{2+})_i$ has been recently documented in single neutrophils or macrophages adhering and spreading on a substrate (Kruskal et al., 1987; Jaconi et al., 1988). It is not clear, however whether the increases in $(Ca^{2+})_i$ were instrumental for cell movement or rather they were a mere consequence of the engagement of surface receptors byion molecules of the substrate.

Table 1. Is a rise in $(Ca^{2+})_i$ required for phagocyte motility?

Cell type	Response	Stimulus	Requirement for $(Ca^{2+})_i$ rise	Reference
Neutrophils[ab]	Phagocytosis	latex particles	no	Hallet & Campbell,1983
Macrophages[a]	Phagocytosis	IgG-coated erythrocytes	yes	Young et al., 1984
Neutrophils[a]	Phagocytosis	IgG-coated yeast	yes/no	Lew et al., 1985
Neutrophils[a]	Phagocytosis	C3b-coated yeast	no	Lew et al., 1985
Macrophages[b]	Phagocytosis	IgG-coated erythrocytes	no	McNeil et al., 1986
Macrophages[b]	Phagocytosis	IgG-coated erythrocytes	no	Di Virgilio et al., 1988a
Neutrophils[a]	Phagocytosis	ConA-coated yeast	no	Rossi et al., 1989
Neutrophils[b]	Phagocytosis	IgG-coated yeast	no	O. Stendahl, unpublished
Neutrophils[b]	Phagocytosis	IgG-coated yeast	no	Della Bianca et al., 1990
Neutrophils[a]	Phagocytosis	C3b-coated yeast	no	Della Bianca et al., 1990
Neutrophils[b]	Chemotaxis	Chemotactic	yes/no	Meshulam et al., 1986
Neutrophils[b]	Chemotaxis	Chemotactic peptide	no	Zigmond et al., 1988
Neutrophils[b]	Secretion	PMA	no	Di Virgilio et al., 1984
Neutrophils[a]	Actin assembly	Chemotactic peptide	no	Sklar et al., 1985
Neutrophils[a]	Actin nucleation	Chemotactic peptide	no	Carson et al., 1986
Neutrophils[a]	Actin assembly	Chemotactic peptide	no	Sha'afi et al., 1986
Neutrophils[a]	Actin assembly	PMA	no	Sheterline et al., 1986

[a] Indicates cell suspensions:

[b] Indicates adherent cells. "yes/no" indicates incomplete inhibition of phagocytosis. Experiments performed with macrophages of different origin (J774 cells or thio glycolate-elicited peritoneal macrophages) and with various procedures for $(Ca^{2+})_i$ monitoring and manipulation are reported. See references for details.

The observation that ingestion of particles can occur at resting $(Ca^{2+})_i$, albeit unexpected, however, is not nearly as surprising as the demonstration that phagocytosis proceeds almost unimpeded in cells drastically depleted of $(Ca^{2+})_i$ (Lew et al., 1985; Di Virgilio et al., 1988a). Two main procedures were used to this aim: first, quin2-loading in the absence of external Ca^{2+} and in the presence of EGTA, and second, EGTA-loading by means of reversible ATP-permeabilization. Both procedures allowed the introduction of substantial Ca^{2+} buffering in the cytosol, thus preventing Ca^{2+} increases and depleting the intracellular Ca^{2+} stores. Although with currently available fluorescent indicators, especially with fura-2, measurements of $(Ca^{2+})_i$ at the lower end of the scale is not as accurate as measurements of $(Ca^{2+})_i$ performed around the kd of the indicator, there are few doubts that phagocytosis still occurs at $(Ca^{2+})_i$ below 20 nM, i. e. 5-10 fold below physiological resting concentrations. Increases in $(Ca^{2+})_i$ seem not to be necessary for locomotion as well. A number of studies indicate that quin2-loaded neutrophils locomote in the absence of extracellular Ca^{2+} or when subjected to extensive $(Ca^{2+})_i$ depletion with Ca^{2+} ionophores in the presence of extra-cellular EGTA (Meshulam et al., 1986; Zigmond et al., 1988).

Several laboratories have investigated the dependence on $(Ca^{2+})_i$ of actin assembly in polymorphonuclear leukocytes stimulated with chemotactic peptides (Sklar et al., 1985; Carson et al., 1986; Sha'afi et al., 1986). In all cases $(Ca^{2+})_i$ was found not to be required for transient actin polymerization (Sklar et al., 1985), actin nucleation (Carson et al., 1986) or increase in cytoskeletal-associated actin (Sha'afi et al., 1986). The kinetics of actin depolymerization, however, was reported to be slower in the absence of $(Ca^{2+})_i$ (Sklar et al., 1985).

If we finally consider also exocytosis of intracellular granules, which involves translocation across the cytoplasm, rearrangement of the cytoskeleton and membrane fusion, the unique role of $(Ca^{2+})_i$ is even more questionable. As initially shown by Pozzan and colleagues (Pozzan et al., 1983; Di Virgilio et al., 1984) a rise in $(Ca^{2+})_i$ is not necessary for secretory exocytosis in human neutrophils. Furthermore, if direct activators of protein kinase C are used, release of secondary granules also occurs at $(Ca^{2+})_i$ below 5-10 nM (Di Virgilio et al., 1984).

Manipulation of $(Ca^{2+})_i$: caveats

All procedures that allow manipulation of $(Ca^{2+})_i$ involve a significant, at times drastic, alteration of the cytoplasmic milieu or of the integrity of the plasma membrane. Loading cells with high (1-2 mM) intracellular concentrations of carboxylate dyes as acetoxymethyl ester conjugates not only chelates intracellular heavy metals besides Ca^{2+}, but also generates four (in the case of quin2) or five (in the case of fura-2) molecules of formaldehyde and acetic acid for each molecule of acetoxymethyl ester conjugate hydrolyzed. Toxic effects on the energy metabolism due to the hydrolysis of intracellular quin2 have been documented (Tiffert et al.,

1984). Depending on the cell type under investigation, it cannot be excluded that intracellular accumulation of these metabolites could be itself inhibitory on the motile machinery.

The other procedures so far used for measuring and clamping $(Ca^{2+})_i$, i. e. fusion with photoprotein-loaded erythrocyte ghosts, scrape loading and ATP-permeabilization, are also traumatic to a variable extent. Scrape-loading and ATP-permeabilization in particular cause lesions in the plasma membrane that allow efflux not only of intracellular ions but also of nucleotides and, in the case of scrape-loading, molecules of MW up to 2,000 kDa (Steinberg et al., 1987; McNeil et al., 1984). It is indeed surprising that after such a harsh treatment macrophages are still capable of phagocytosis (McNeil et al., 1986; Di Virgilio et al., 1988a). Furthermore prolonged deprivation of $(Ca^{2+})_i$ can be harmful to other cellular functions and affect the motile response indirectly. We have preliminary observations indicating that EGTA-loaded mouse macrophages undergo drastic and irreversible alterations in plasma membrane permeability within 2 to 3 h of EGTA-loading by means of ATP-permeabilization (Picello and Di Virgilio, manuscript in preparation). Permeabilization of the plasma membrane as a consequence of $(Ca^{2+})_i$ deprivation is not surprising given that an established procedure for skinning skeletal muscle fibers involves overnight incubation of freshly dissected fibers in the presence of high EGTA concentrations (Wood et al., 1975). Therefore in case suppression of locomotion, phagocytosis or exocytosis by $(Ca^{2+})_i$-depleting treatment is observed, it is always crucial to check that:

1) the motile response under investigation is conserved in control cells, quin2-loaded in the presence of physiological extracellular Ca^{2+} concentration (and therefore not $(Ca^{2+})_i$ depleted);

2) the inhibition is reversible, i. e. the phagocytes regain normal motility once the incubation medium is resupplemented with Ca^{2+}.

In case ATP-permeabilization is chosen for loading Ca^{2+} buffers into the cytoplasm, is mandatory:

1) to verify that perfect resealing is achieved after the permeabilization step;

2) to allow a suitable interval (at least 10-15 min in the case of mouse macrophages) for cell recovery after the ATP treatment.

Recent observations on the intracellular redistribution of carboxylate dyes in macrophages have raised another relevant problem for the interpretation of experiments aimed at investigating the role of $(Ca^{2+})_i$ in motile responses. We have discovered that mouse macrophages possess organic-anion transporters that remove fluorescent dyes, including fura-2 and, albeit to a lesser extent, quin2, from the cytoplasmic matrix of these cells (Di Virgilio et al., 1988b). The dyes are sequestered within cytoplasmic vacuoles and secreted into the extracellular medium. It is all too obvious that under these conditions the real cytoplasmic concentrations of the Ca^{2+} buffers, and therefore chelation of $(Ca^{2+})_i$, may be much less than expected. It is mandatory in this case to measure the cytoplasmic concentration and distribution of the dyes and to adopt procedures aimed at preventing the dyes from leaving the cytoplasm. In our hands, inhibitors of organic-anion transport, such as probenecid and sulphinpyrazone, proved to be useful to prevent fura-2 leakage and sequestration in macrophages and other cells and facilitated measuring of $(Ca^{2+})_i$ (Di Virgilio et al., 1988b; Di Virgilio et al., 1988c).

The phosphoinositide cycle and actin assembly

If we admit that the cytoskeleton may be under multiple control, it is possible that $(Ca^{2+})_i$, although not required, contributes to the regulation of phagocyte motility by other factors. Evidence is accruing in favor of a role for the phosphoinositide cycle in the modulation of gelsolin-actin interaction (Stossel, 1988). According to this view hydrolysis of $PtdIns4,5P_2$ triggered by chemoattractants causes generation of $Ins1,4,5P_3$ and of DAG. $Ins1,4,5P_3$ releases Ca^{2+} from the intracellular Ca^{2+} stores, possibly the newly discovered organelles named "calciosomes" (Volpe et al., 1988). In the presence of a raised $(Ca^{2+})_i$ gelsolin complexes actin. $PtdIns4,5P_2$ resynthesis, in synergism with the extrusion of Ca^{2+} from the cytoplasm, reverses the interaction of gelsolin with actin. This fascinating explanation of phagocyte contractility, however, might be invalidated by the demonstration that drastic depletion of $(Ca^{2+})_i$ has little effect on cell motility although it suppresses $Ins1,4,5P_3$ accumulation (and therefore $PtdIns4,5P_2$ hydrolysis and resynthesis) in human neutrophils (Lew et al., 1986). In addition very recently direct demonstration was provided that actin assembles in the absence of $PtdIns4,5P_2$ turnover in human neutrophils stimulated with the chemotactic peptide fMet-Leu-Phe. These observations, although not excluding the possibility of a regulatory role for products of the phosphoinositide cycle in actin assembly, nonetheless leave open the quest for the main regulatory mechanism of leukocyte cytoskeleton contractility.

Conclusion

Phagocyte motility has become the point of intersection of several interrelated fields: structure and function of surface receptors, molecular biology of the cytoskeleton, biochemistry of the calcium and phosphoinositide transducing systems. Techniques allowing genetic modification of key cytoskeletal proteins or access to and manipulation of the cytoplasm have so far provided invaluable (and unexpected) insights on the mechanism of the motile response. The naive dream of the search for "<u>the regulator</u>" of cell motility must be reexamined, and this may lead to the possible discovery of multiple, and even redundant regulatory processes affecting the cytoskeleton (Zigmond, 1988).

ACKNOWLEDGEMENTS

This work was supported by grants from MPI (40 % and 60 %), from CNR (Projects Oncology and Biotechnology and Bioinstrumentation), and from AICR. The authors are indebted to Prof. T. Pozzan and Dr. F. Michelangeli for helpful comments.

REFERENCES

Adams, R.J., Pollard, T.D., 1989. Binding of myosin I to membrane lipids. Nature (London) 340:565.

Brundage, R.A., Fay, F.S., 1989. The role of Ca^{2+} in the polarization and chemotaxis of newt eosinophils. J. Cell Biol. 107:17a (abstract).

Bengtsson, T., Rundquist, I., Stendahl, O., Wymann, M., Andersson, T., 1988. Increased breakdown of phosphatidyl-inositol 4,5-bisphosphate is not an initiating factor for actin assembly in human neutrophils. J. Cell Biol. 263:17385.

Campbell, A.K., Hallet, M.B., 1983. Measurement of intracellular calcium ions and oxygen radicals in polymorphonuclear leukocyte-erythrocyte ghost hybrids. J. Physiol. (London) 338:537.

Carson, M., Weber, A, Zigmond, S.H., 1986. An actin-nucleating activity in polymorphonuclear leukocytes is modulated by chemotactic peptides. J. Cell Biol. 103:2707.

Della Bianca, V., Grzeskowiak, M., Rossi, F., 1990. Studies of molecular regulation of phagocytosis and activation of NADPH oxidase in neutrophils. IgG and C3b-mediated ingestion and associated respiratory burst independent of phosphatidyl turnover and Ca^{2+} transients. J. Immunol. in press.

De Lozanne, A., Spudich, J.A., 1987. Disruption of the Dictyostelium myosin heavy chain gene by homologous recombination. Science 236:1086.

Di Virgilio, F., Lew, D.P., Pozzan, T., 1984. Protein kinase C activation of physiological processes in human neutrophils at vanishingly small cytoslic Ca^{2+} levels. Nature (London) 310:691.

Di Virgilio, F., Meyer, B.C., Greenberg, S., Silverstein, S.C., 1988a. Fc-receptor-mediated phagocytosis occurs in macropha-ges at exceedingly low cytosolic Ca^{2+} levels. J. Cell Biol. 106:657.

Di Virgilio, F., Steinberg, T.H., Swanson, J.A., Silverstein, S.C., 1988b. Fura-2 secretion and sequestration in macrophages. A blocker of organic anion transport reveals that these processes occur via a membrane transport system for organic anions. J. Immunol. 140:915.

Di Virgilio, F., Fasolato, C., Steinberg, T.H., 1988c. Inhibitors of membrane transport system for organic anions block fura-2 excretion from PC12 and N2A cells. Biochem. J. 256:959 .

Di Virgilio, F. Stendahl, O., Pittet, D., Lew D.P., Pozzan, T., 1990. Cytoplasmic calcium in phagocyte activation. Current Topics in Membranes and Transport 35 (in press).

Fukui, Y., Lynch, T.J., Brzeska, H., Korn, E.D., 1989. Myosin I is located at the leading edges of locomoting

Dictyostelium amoebae. <u>Nature</u> (London) 341:328.

Griffin, F.M., Jr., Griffin, J.A., Silverstein, S.C., 1976. Studies on the mechanism of phagocytosis. II. The interaction of macrophages with anti-immunoglobulin IgG-coated bone marrow-derived lymphocytes. <u>J. Exp. Med.</u> 144:788.

Jaconi, M.E.E., Rivest, R.W., Schlegel, W., Wollheim, C.B., Pittet, D., Lew, P.D., 1988. Spontaneous and chemoattractant-induced oscillations of cytosolic free Ca^{2+} in single adherent human neutrophils. <u>J. Biol. Chem.</u> 263:10557.

Janmey, P.A., Stossel, T.P., 1987. Modulation of gelsolin function by phosphatidylinositol 4,5-bisphosphate. <u>Nature</u> (London) 325:362.

Knecht, D.A., Loomis, W.F., 1987. Antisense RNA inactivation of myosin heavy chain gene expression in Dictyodtelium Discoideum. <u>Science</u> 236:1081.

Kruskal, B.A., Maxfield, F.R., 1987. Cytosolic free calcium increases before and oscillates during frustrated phagocytosis in macrophages. <u>J. Cell Biol.</u> 105:2685.

Lassing, I., Lindberg, U., 1985. Specific interaction between phosphatidylinositol 4,5-bisphosphate and profilactin. <u>Nature</u> (London) 314:472.

Lew, P.D., Andersson, T., Di Virgilio, F., Pozzan, T., Stendahl, O., 1985. Ca^{2+}-independent phagocytosis in human neutrophils. <u>Nature</u> (London) 315:509.

Lew, P.D., Monod, A., Krause, K.H., Waldvogel, F.A., Biden, T.J., Schlegel, W., 1986. The role of cytosolic calcium in the generation of inositol 1,4,5-trisphosphate and inositol 1,3,4-trisphosphate in HL60 cells: differential effects of chemotactic peptide receptor stimulation at distinct Ca^{2+} levels. <u>J. Biol. Chem.</u> 261:13121.

McNeil, P.L., Murphy, R.F., Lanni, F., Taylor, D.L., 1984. A method for incorporating macromolecules into adherent cells. <u>J. Cell Biol.</u> 98:1556.

McNeil, P.L. Swanson, J.A., Wright, S.D., Silverstein, S.C., Taylor, D.L., 1986. Fc-receptor-mediated phagocytosis occurs in macrophages without an increase in average $(Ca^{2+})_i$. <u>J. Cell Biol.</u> 102:1586.

Meshulam, T., Proto, P., Diamond, R.D., Melnick, D.A., 1986. Calcium modulation and chemotactic response: divergent stimulation of neutrophil chemotaxis and cytosolic calcium response by the chemotactic peptide receptor. <u>J. Immunol.</u>137:1954.

Pozzan, T., Lew, D.P., Wollheim, C.B., Tsien, R.Y., 1983. Is cytosolic ionized calcium regulating neutrophil activation? <u>Science</u> 221:1413.

Rossi, F., Della Bianca, V., Grzeskowiak, M., Bazzoni, F., 1989. Studies on molecular regulation of phagocytosis in neutrophils. Con A-mediated ingestion and associated respiratory burst independent of phosphoinositide turnover, rise in $(Ca^{2+})_i$ and arachidonic acid release. <u>J. Immunol.</u> 142:1625 .

Sawyer, D.W., Sullivan, J.A., Mandell, G.L., 1985. Intracellular free calcium localization in neutrophils during phagoytosis. <u>Science</u> 230:663.

Sha'afi, R.I., Shefcyk, J. Yassin, R., Molski, T.F.P., Volpi, M., Naccache, P.H., White, J.R., Feinstein, M.B., Becker, E.L., 1986. Is a rise in intracellular concentration of free calcium necessary or sufficient for stimulated cytoskeletal-associated actin. J. Cell Biol. 102:1459.

Sheterline, P. Rickard, J.E., Richards, R.C., 1984. Fc-receptor-directed phagocytic stimuli induce transient actin assembly at an early stage of phagocytosis in neutrophil leukocytes. Eur. J. Cell Biol. 34:80.

Sheterline, P., Rickard, J.E., Boothroyd, B., Richards, R.C., 1986. Phorbol esters induce rapid actin assembly in neutrophil leukocytes independently of changes in $(Ca^{2+})_i$ and pH_i. J. Muscle Res. Cell Motil. 7:405.

Sklar, L.A., Omann, G.M., Painter, R.G., 1985. Relationship of actin polymerization and depolymerization to light scattering in human neutrophils: dependence on receptor occupancy and intracellular Ca^{2+}. J. Cell Biol. 102:1459.

Silverstein, S.C., Steinmann, R.M., Cohn, Z.A., 1977. Endocytosis. Annu. Rev. Biochem. 46:669.

Silverstein, S.C., Greenberg, S., Di Virgilio, F., Steinberg, T.H., 1989. Phagocytosis. in "Fundamental Immunology", W. Paul ed. Raven Press, New York.

Steinberg, T.H., Newman, A., Swanson, J.A., Silverstein, S.C., 1987. ATP^{4-} permeabilizes the plasma membrane of mouse macrophages to fluorescent dyes. J. Biol. Chem. 262:8884.

Stossel, T.P., 1988. The mechanical responses of withe blood cells. in "Inflammation: Basic Principles and Clinical Correlates". Gallin, J.I., Goldstein, I.M. and Snydermann R. eds. Raven Press, Ltd. New York.

Tiffert, T., Garcia-Sancho, J., Lew, V.L., 1984. Irreversible ATP depletion caused by low concentrations of formaldehyde and of calcium-chelator esters in intact human red cells. Biochim. Biophys. Acta. 773:143.

Trotter, J.A., Adelstein, R.S., 1979. Macrophage myosin: regulation of actin-activated ATPase activity by phosphorylation of the 20,000-dalton light chain. J. Biol. Chem. 260:8781.

Volpe, P., Krause, K.H., Hashimoto, S., Zorzato, F., Pozzan T., Meldolesi, J., Lew, D.P., 1988. Calciosome, a cytoplasmic organelle: the inositol 1,4,5-trisphosphate-sensitive Ca^{2+} store of non-muscle cells. Proc. Natl. Acad. Sci. USA 85:1091.

Wang, E., Michl., J., Pfeffer, L.M., Silverstein, S.C., 1984. Interferon suppresses pinocytosis but stimulates phagocytes in mouse peritoneal macrophages: related changes in cytoskeletal organization. J. Cell Biol. 98:1328.

Wood, D.S., Zollman, J.R., Reuben, J.P, Brandt, P.W., 1975. Human skeletal muscle: properties of the chemically skinned fiber. Science 187:1075.

Wuestenhube, L.J., Luna, E.J., 1987. F-actin binds to the cytoplasmatic surface of ponticulin, a 17-kDa integral glycoprotein from Dictyostelium discoideum plasma membranes. J. Cell Biol. 105:1741.

Young, J.D.-E., Ko, S.S., Cohn, Z.A., 1984. The increase in intracellular free Ca^{2+} associated with IgG 2b/1 Fc receptor-ligand interaction: role in phagocytosis. Proc. Natl. Acad. Sci. USA 81:5430.

Zigmond, S.H., Slonczewski, J.L., Wilde, M.W., Carson, M., 1988. Polymorphonuclear leukocyte locomotion is insensitive to lowered cytoplasmic calcium levels. Cell Motility and the Cytoskeleton 9:184.

DIVERSITY IN MOTILE RESPONSES OF HUMAN NEUTROPHIL GRANULOCYTES: FUNCTIONAL MEANING AND CYTOSKELETAL BASIS

Hansuli Keller, Verena Niggli, Arthur Zimmermann

University of Bern / Institute of Pathology
Freiburgstrasse 30
3010 B e r n / Switzerland

INTRODUCTION

Neutrophil granulocytes are multifunctional cells, capable of locomotion, chemotaxis, adhesion, pinocytosis, phagocytosis, intracellular killing or degradation and exocytosis. Several of the functions require generation of force and may thus be associated with different forms of motility. In the circulating blood of healthy individuals neutrophils are in a relatively quiescent state, i.e. they are spherical and nonmotile. Activation and regulation of neutrophil functions can, to some extent, occur in a selective manner. Most agonists activate only some neutrophil functions but not others. Or they stimulate at least some functions to a much greater extent than others. Chemotaxis for example is an important neutrophil function, but only few agonists are actually chemotactic, while many others are not. Furthermore an agonist, e.g. a chemotactic factor, may elicit pinocytosis or chemotaxis at lower concentrations than exocytosis (Davis et al., 1986). Such qualitative, quantitative and temporal differences may help to understand the relationship, if any, between these functions. Thus, the F-actin peak preceeds the time maximum for pinocytosis (Davis et al., 1986), and the maximal pinocytotic response occurs at an earlier time point than full development of polarity (Fig. 1) and locomotion.

For many decades, shape changes in leukocytes have been mainly associated with locomotor activity. More recent studies show a great diversity of motile responses in neutrophils. It is well established that locomoting leukocytes show a polarized triangular morphology which is so characteristic that it is sometimes also called locomotor morphology. This has for a long time been considered as the motile response of neutrophils to the extent that leukocyte motility has been equated with locomotor activity. More recent studies show that there are other forms of motility in neutrophils, which may be associated with other functions rather than with locomotion. Therefore, it may be more appropriate to equate motility with shape changes in general rather than just with the one form characteristic for locomotion. Thus, if we are refering to motility in the present article, we mean any type of shape changes and not necessarily locomotion.

Numerous other forms of motility, i.e. different types of shape changes can, to some extent, be elicited by different agonists (Roos et al., 1987; Zimmermann, Keller, Cottier, 1988; Zimmermann, Gehr, Keller, 1988; Robinson et al., 1987; Keller et al., subm.). The shape produced by different agents can be so distinct that one can determine the agonist by its morphological effects. This is an interesting addition to previous studies showing that different agonists can activate different neutrophil functions such as chemokinesis, chemotaxis, phagocytosis, pinocytosis, exocytosis, production of oxygen radicals and others because shape changes and function may correlate to some extent. Polarized leukocytes have the capacity to locomote efficiently, whereas spherical or non-polar cells do not (Lewis, 1934; Zigmond and Sullivan, 1979; Zigmond et al., 1981; Keller, 1983; Roos et al., 1987; Zimmermann, Keller, Cottier, 1988; Zimmermann, Gehr, Keller, 1988; Keller et al., subm.). Also on the level of a single cell the shape change has functional consequences. The motor of the cells is at the front (Keller and Bessis, 1975), whereas internalization by receptor-mediated endocytosis occurs preferentially at the rear end of the neutrophil and the cytoskeleton is rearranged accordingly (Davis et al., 1982). The present article summarizes the present knowledge on different types of neutrophil motility and addresses the question how different types of shape changes may be related to different types of neutrophil functions requiring motor activity such as locomotion, chemotaxis, phagocytosis or pinocytosis.

Cytoskeletal changes play a major role in controlling motility, though the precise mechanisms are not fully understood. In the present article we are also going to address the question whether different forms of cytoskeletal organization are associated with different types of shape changes. We hope that this systematic approach will provide useful information on the relationship between shape and the functional state of cells, and improve our understanding of the mechanisms controlling cell motility.

DISTINCT TYPES OF SHAPE CHANGES INDUCED BY DIFFERENT TYPES OF AGONISTS

Shape of unstimulated neutrophils. Normal neutrophils from the peripheral human blood are unstimulated and, therefore, spherical and nonmotile (Fig. 1). Shape changes are a fairly sensitive indicator of leukocyte activation. Therefore, we postulate that any studies on leuocyte activation should be conducted with initially spherical cells. This applies in particular to studies on motility, because some shape change responses (e.g. blebbing in response to microtubule-disassembling agents) may not be detectable if the cells are already activated in some other way, for instance by chemotactic peptides (Keller et al., 1984). Another example is that polarity can no longer be induced with chemotactic peptides in cells which have been stimulated with PMA or diacylglycerols before. The PMA-induced shape changes are dominant over the ones induced by chemotactic peptides (Roos et al., 1987). This

means that the characteristics of these shape changes induced
by particular types of agonists can only be properly described
if the control cells are not stimulated spontaneously or by
chemicals. If initially spherical and nonmotile neutrophils
are stimulated with different classes of agonists, they can
show distinct types of shape changes, which are often so
characteristic that mere morphological examination of the
cells may allow the observer to diagnose the type of stimulant
used (Lewis, 1934; Roos et al., 1987; Zimmermann, Keller,
Cottier, 1988; Zimmermann Gehr, Keller, 1988; Keller et al.,
subm.). However, it is possible that at certain points in the
time course of the responses or at certain concentrations of
a given agonist these characteristics may not be developed
(Zigmond and Sullivan, 1979; Keller, 1983; Keller et al.,
subm.) or that overstimulated (deactivated) neutrophils become
spherical again (Roos et al., 1987; Zimmermann, Gehr, Keller,
1988).

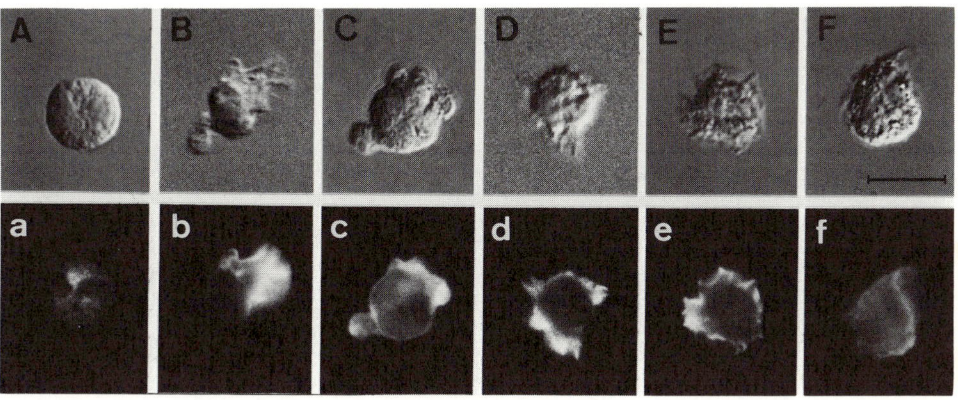

Fig. 1. Neutrophil shape and redistribution of F-actin.
Examples of unstimulated neutrophils (A,a), polarized (front
oriented towards the top) neutrophils in 10-9M fNLPNTL (B,b),
10-5M colchicine (C,c), or nonpolar cells with surface
projections in 10-9M PMA (D,d), 3.10-4 OAG (E,e) or in D2O
(F,f). Cells incubated for 30 minutes, fixed with paraformal-
dehyde and stained for F-actin with NBD-phallacidin were
photographed with DIC-optics (top) or fluorescence microscopy
(bottom). Bar: 10μm.

Front-tail polarity. The best known shape change response
of leukocytes is front-tail polarity (Fig. 1). This is the
characteristic shape of locomoting cells (Lewis, 1934). Chemotactic factors have been found capable to induce front-tail
polarity, continuous cytoplasmic streaming towards the front,
and stimulated locomotion. Therefore, it has been proposed to
use the polarization assay as an assay for chemotaxis (Haston
and Shields, 1985). Though this may be a useful and convenient
screening assay, it is not fully reliable. First, at the early
stage of the response (up to about 1-2 minutes) neutrophils
stimulated with chemotactic peptides show ruffling all over
the surface but no front-tail polarity and no locomotor
activity (Davis et al., 1982; Zigmond and Sullivan, 1979). The

polarisation response to increasing peptide concentration is biphasic. High cytotaxin concentrations decrease polarity and generate nonpolar cells with surface projections (Keller, 1983). Second and more important, some agents like microtubule-disassembling agents (Keller et al., 1984) or the protein kinase C inhibitor H-7 (Keller et al., subm.), which have chemokinetic but not chemotactic properties (for definitions see Keller et al., 1977), may elicit some type of polarity. In contrast to chemotactic peptides, cells treated with microtubule-disassembling drugs show blebbing rather than ruffling at the leading front. A peculiar type of polarity has been found in cells stimulated with H-7. A proportion of these cells may develop polarity characterized by the absence of a tail-knob. These cells can be very elongated and show ruffling at the leading front (Keller et al., subm.). Thus a polarisation assay is a more accurate measure for chemokinetic than for chemotactic activity.

Nonpolar cells with surface projections and vacuole formation. This type of shape change is found after neutrophil stimulation with active phorbol esters, in particular PMA (Roos et al., 1987) or with diacylglycerols such as OAG, diC8, and diC10 (Zimmermann et al., 1988). Similar responses have been seen in lymphocytes (Keller et al., 1989). In contrast to polarized cells these nonpolar neutrophils with surface projections show no continuing expansion of pseudopods and continuing cytoplasmic streaming into one direction only. Instead, there are ruffles distributed more or less all over the surface and streaming is not well-defined (Roos et al., 1987). Numerous invaginations and intracellular vacuoles are prominent and are characteristic features of these cells (White and Estensen, 1974; Roos et al., 1987; Robinson et al., 1987). The term "nonpolar cells" emphasizes the absence of net cell polarity and that protrusions are again withdrawn after a short time and, therefore, do not result in significant behavioral polarity, e.g. locomotion.

Other forms of nonpolar cells with surface projections. Nonpolar cells with surface projections characterized by very short ruffles appearing all over the surface are observed shortly (e.g. within about one minute) after stimulation with chemotactic peptides or the protein kinase inhibitor H-7. At this stage, morphological responses to these two agonists are not distinguishable (Keller et al., subm.). Later, polarity develops gradually to a greater (chemotactic peptides) or lesser (H-7) extent. The cell length varies with the concentration of chemotactic peptides and at very high concentrations of chemotactic peptides (e.g. 10-6 M fMNLP) (Keller, 1983; Haston and Shields, 1985) or of H-7 (300 μM) (Keller et al., subm.), the cells become shorter again, polarity is lost and there is an increase in nonpolar cells with surface projections.

A somewhat different motile response occurs in the presence of D2O (Zimmermann et al., 1988). The shape changes are less pronounced because they are very slow and because the projections are shorter. In addition to numerous very short projections there is often one relatively large lamella. At least in some of these cells the lamella is making a kind of circus movement, indicating what may be called circular polarity.

In summary, the data available so far show that neutrophils can respond to different classes of agonists by fairly distinct types of shape changes. We suspect that testing more agents systematically for their capacity to induce shape changes will lead to an even greater variety of responses than those known so far. It becomes increasingly important to understand their functional significance.

RELATIONSHIP BETWEEN AGONIST-INDUCED SHAPE CHANGES AND FUNCTIONAL ACTIVATION

Chemotactic factors. It is now clear that different agonists activate different functions. Chemotaxis is only induced by a well-defined group of agonists (cytotaxins, chemotactic factors). Chemotactic factors have also chemokinetic activity. This has been shown for oligopeptides (Keller, 1983; Keller et al., 1983), LTB4 (Ford-Hutchinson et al., 1980; Evans et al., 1987), HETEs (Evans et al., 1987), which are all capable to stimulate the locomotor activity of neutrophils. The locomotor activity correlates with the extent of cell elongation. The longer the axis the higher is the locomotor activity (Keller, Zimmermann, Cottier, 1983; Haston and Shields, 1985). Behavioral polarity becomes established with high probability once a pseudopod of 3-5 μm is formed (Zigmond et al., 1981). The capacity of chemotactic factors to stimulate exocytosis has been tested extensively under conditions, where these peptides induce random or directional locomotion, i.e. in absence of cytochalasin B. Under these conditions the capacity of chemotactic factors to induce exocytosis is relatively small or absent (Rollins et al., 1983). They are also capable of inducing pinocytosis (Davis et al., 1982.; Daukas et al., 1983; Robinson et al., 1987). However, the net uptake is small as compared to PMA-stimulated cells (Keller and Zimmermann, 1987) (Fig. 2). Many structural and functional changes are associated with this type of shape change on the level of single cells. The motor with its components becomes mainly located at the front, coated pits and vesicles instrumental in pinocytosis are mainly at the rear end. Also surface receptors are redistributed (for review see Keller and Zimmermann, 1987).

Microtubule-disassembling agents such as colchicine, nocodazole and vinblastine have chemokinetic properties, even though they lack chemotactic activity. Depending on the initial locomotor activity of the control cells, these agents may have no effect or stimulate or inhibit neutrophil locomotion. Locomotion is stimulated provided the control cells are spherical and non-motile (Keller et al., 1984).

Another agent which has chemokinetic but not chemotactic properties is the protein kinase inhibitor H-7. Depending on the conditions, it may stimulate or inhibit locomotion of neutrophils. It can also suppress neutrophil polarity induced by chemotactic peptides and accordingly inhibit peptide-induced locomotion. It has a small stimulating effect on the net uptake of fluorescent dextran (Keller et al., subm.).

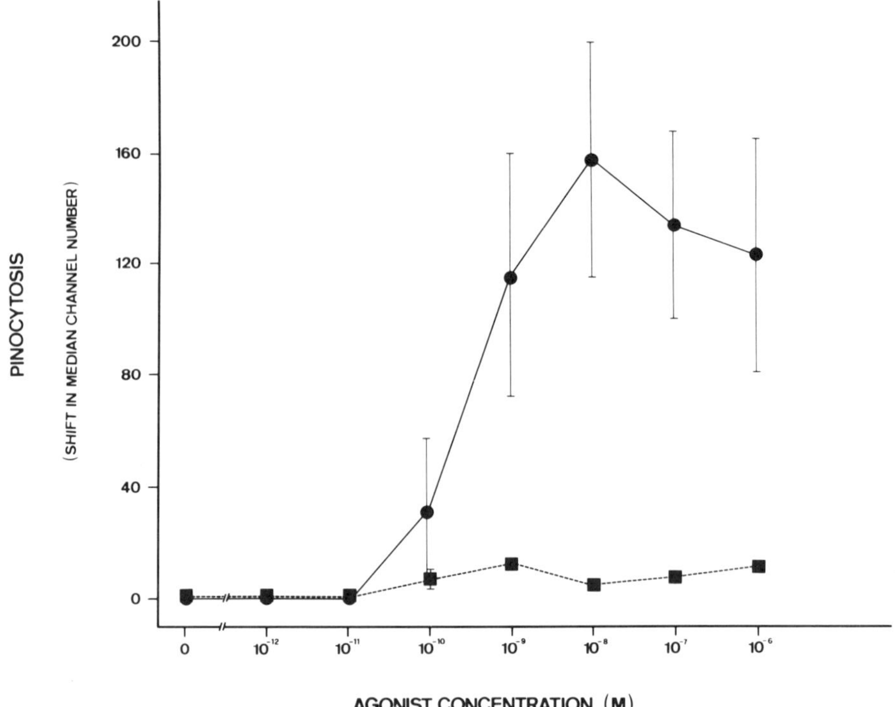

Fig. 2. Differential effects of fNLPNTL vs PMA on pino-
cytosis.The net uptake of FITC-dextran (20mg/ml) in the
presence of fNLPNTL (■-------■) or PMA (●———●) using 2% HSA-
Gey's-Hepes without added Ca++ and Mg++ and an incubation time
of 30 minutes. The results of flow cytometry analysis are
expressed as increase in median channel number over unstimulated controls. Specific uptake of FITC-dextran by controls
was 9.5 channels. 3 experiments ± SDM.

PMA and diacylglycerols. Most studies in neutrophils and
lymphocytes show that short-term stimulation with PMA or diacylglycerols inhibits random and directional locomotion
(Gallin et al., 1978; Hoult and Nourshargh, 1985; Noursharg
and Hoult, 1987; Roos et al., 1987; Keller et al., 1989;
Zimmermann et al., 1988), and that they are not chemotactic
(Cox et al., 1986). One study based on the under-agarose
assay, an indirect method, suggests that 1,2-diacylglycerol
is chemotactic (Wright et al., 1988). No chemotactic activity
could be found in a direct visual assay using micropipettes
(Roos et al., 1987). However, PMA (Keller and Zimmermann,
1987) and diacylglycerols (Keller, manuscript in preparation)
produce a very marked stimulation of pinocytosis as measured
by the net uptake of FITC-dextran (Fig. 2). Furthermore, it is
well-established that PMA and diacylglycerols are potent
stimulators of exocytosis, in particular secondary granule
secretion (White and Estensen, 1974); Gallin et al., 1978;
Robinson, 1987; Cox et al., 1986; White et al., 1984, Fletcher

et al., 1982). There is also evidence that pinocytosis and exocytosis may be linked (receptor recycling).

D2O. D2O inhibits locomotion and lacks chemotactic activity (Zimmermann et al., 1988). It stimulates pinocytosis to a small extent (Keller, to be published) and the effect on exocytosis has not been studied as yet. The function of circus movement is not known.

THE ROLE OF THE NEUTROPHIL CYTOSKELETON IN SHAPE CHANGES AND LOCOMOTION

The composition of the neutrophil cytoskeleton. Neutrophils contain the three major cytoskeletal systems: microfilaments, microtubules and the intermediate type filaments (Parysch and Eckert, 1984; Devreotes and Zigmond, 1988). Several actin-associated proteins have already been identified in neutrophils. Myosin and actin-binding protein have been located in the anterior pseudopods of cells crawling towards yeast particles (Valerius et al., 1981). Other proteins involved in regulating the extent of actin polymerization, such as gelsolin, and protein 4.1, a component of the erythrocyte membrane skeleton, have been identified in neutrophils (Omann et al., 1987; Spiegel et al., 1984). Fodrin has been detected in purified neutrophil plasma membrane fractions (Jesaitis et al., 1988). Recently, the presence of the actin cross-linking and bundling protein α-actinin, and of the putative actin-membrane linker protein vinculin, have been demonstrated in human neutrophils (Niggli and Jenni, 1989).
The exact role of these cytoskeletal components in neutrophil locomotion is yet to be understood.

Different types of shape are associated with different types of cytoskeletal organization.

- Resting cells: In spherical unstimulated cells F-actin is diffusely distributed. Only a fraction of the total cellular actin is in the polymerized form, as measured by assaying the amount of Triton X-100-insoluble actin (Omann et al., 1987) (Table 1). Microtubules are arranged radially in resting cells, originating from a centrally located single microtubule organizing center (Anderson et al., 1982).

- Cells with front-tail polarity: Chemotactic peptides induce a rapid and marked increase in neutrophil F-actin, which correlates with the assembly of actin into a Triton X-100-insoluble network (Omann et al., 1987) (Table 1). This rapid increase is followed by a slow partial depolymerization to a lower level. In our hands, the amount of F-actin (see Fig. 1) or the amount of cytoskeleton-associated actin is still significantly increased above controls after longer periods (up to 30 min) of stimulation (Niggli and Jenni, 1989, and Table 1). Our data agree with previously published data, which were obtained with the DNase I inhibition assay (Fechtheimer and Zigmond, 1983; Rao and Varani, 1982), a measure for G-actin, or the nitrobenzoxadiazole (NBD) phallacidin extraction assay, a measure for F-actin (Howard and Oresajo, 1985). In contrast to these findings, cytoskeletal actin was reported to return to basal levels within 30-120 seconds of stimulation (White et al., 1982). In these experi-

ments the same technique was used as in our investigations.
The reason for this discrepancy is not known. The point is of
importance, as front-tail polarity and locomotion occur only
in the later phase of activation, when a decrease in the
initial high level of cytoskeletal actin has taken place
(Table 1). It is therefore not clear, if locomotion is
associated with a net increase in F-actin or not.

In addition it has been shown that chemotactic stimulation of neutrophils leads to a rapid translocation of α-acti-
nin into the cytoskeleton (Niggli and Jenni, 1989, and Table
1). This rearrangement accompanies the formation of the actin
network. It may be an important event in pseudopod extension.

Table 1. The effect of chemotactic peptide on neutrophil
shape and cytoskeletal actin

Experimental	Cytoskeletal actin (% of total)	spherical cells including cells with unifocal projections (% of total)	non-polar cells with surface projections (% of total)	cells with frontail polarity (% of total)
Experiment 1				
- Medium, 1 min, 37 ^{O}C	7.3 %	68 %	14 %	18 %
- fNLPNTL, 10^{-8}M, 1 min, 37 ^{O}C	48.7 %	10 %	86 %	4 %
Experiment 2				
- Medium, 30 min, 37 ^{O}C	7.2 %	87 %	3 %	10 %
- fNLPNTL, 10^{-8}M, 30 min, 37 ^{O}C	18.4 %	12 %	6 %	82 %

[a] The isolated neutrophils were preincubated with 10 mM EDTA at 37 ^{O}C,
followed, by addition of medium or peptide, and a further incubation,
as indicated. Aliquots of the same population of cells were used for
assessment of morphology and determination of cytoskeletal actin (White
et al. 1982). Cytoskeletal actin is given as the mean of duplicates.

Cells with established polarity show a specific location
of F-actin: the latter is enriched in the leading lamellae and
to a lesser extent in the tail knob of the locomoting cells
(see Fig. 1). As outlined above, myosin and actin binding
protein also show a polar location in locomoting neutrophils
(Valerius et al., 1981). Chemotactic stimulation with f-Met-
Leu-Phe also affects the organization of microtubules. An
elongation of microtubules occurs parallel to the direction of
cell migration, both in chemotactic gradients and in uniform
concentrations of stimuli (Anderson et al., 1982). According

to these authors, no significant changes in microtubule number per cell occurred during chemotactic activation. Chemotactic stimulation of neutrophils by C5A or f-Met-Leu-Phe results in transient centrosome splitting into two solitary centrioles surrounded by asters of microtubules (Schliwa et al., 1982). This process has only been observed for attached cells, not for cells in suspension, which can polarize nevertheless. Schliwa et al. (1982) also find that stimulation of attached cells is accompanied by a small increase in the average length of the microtubules.

Vimentin filaments appear to be bundled in the uropod of f-Met-Leu-Phe-treated neutrophils (Parysek and Eckert, 1984). The authors suggest that the tail could be made rigid by an accumulation of filaments, lending stability to the asymmetric shape of the motile cell.

<u>Nonpolar cells with surface projections</u>: The agents H-7 and active phorbol esters which induce a nonpolar shape with surface projections also increase significantly the amount of cytoskeleton-associated actin (Sha'afi et al., 1983; Sheterline et al., 1986; Keller et al., subm.). The extent of the rise in cytoskeletal actin induced by 300 μM H-7 after 30 minutes is comparable to that induced by chemotactic peptide after 1 minute of stimulation (Keller et al., 1989, subm.). The increase in cytoskeletal actin induced by phorbol ester is smaller than that observed at 1 minute after addition of chemotactic peptides (Sha'afi and Molski, 1987). Its time-course is, in contrast to that induced by peptide, not biphasic, and the increase persists at longer times of stimulation (Sha'afi and Molski, 1987; Sheterline et al., 1986). At 30 minutes after addition of 10-8M phorbol myristate acetate the increase in cytoskeletal actin is, in our hands, somewhat higher than that induced by chemotactic peptide at this time point, but it is lower than the peptide-induced increase at 1 minute (V. Niggli, unpublished observations). Phorbol myristate acetate , comparable to chemotactic peptide, has also been found to stimulate incorporation of α-actinin into the cytoskeleton, in parallel with actin (V. Niggli, unpublished observations). Both H-7 (Keller et al., subm.) and phorbol ester (Fig. 1) induce an enrichment of F-actin in the surface protrusions.

Phorbol ester, similar to chemotactic peptides, induces centrosome splitting in neutrophils. Moreover, the agent increases the total number of microtubules associated with the centrosome, and also increases overall polymer length (Schliwa et al., 1983). Actin may be involved in the positioning and motility of centrosomes (Euteneuer and Schliwa, 1985). This could explain, why chemotactic peptide and phorbol ester, agonists which both increase F-actin, also both induce centrosome splitting.

Relation between cytoskeletal rearrangement and function. As outlined above, shape changes induced by various agents are accompagnied by changes in the organization of actin filaments, microtubules and intermediate filaments. We will now address the question on the functional contribution of these cytoskeletal rearrangements to motility and migration. Motility, locomotion and chemotaxis are certainly dependent on a functional actin cytoskeleton, as cytochalasin which caps

actin filaments and inhibits further filament growth, abo-
lishes pseudopod formation, motility and locomotion (Keller et
al., 1984; Omann et al., 1987). The drug inhibits the chemotactic peptide-induced increase in cytoskeletal actin, and
also reduces the residual cytoskeletal actin to a lower level
(White et al., 1983). We conclude that cytochalasin interferes
with the dynamic pool of the actin filaments. Microtubules are
not necessary for locomotion, as anucleate, microtubule and
centrosome-free neutrophil fragments are capable of directed
migration and phagocytosis (Malawista, 1986). Microtubules
may however serve to stabilize and orient the nucleus.
Nothing is yet known on the role of vimentin in motile
processes of neutrophils.

The dynamic actin network appears thus to be an essential
component of the motor of the neutrophils. According to the
findings summarized above, all agents that induce shape
changes, ruffling, motility and locomotion also induce a shift
of F-actin into surface protrusions. This redistribution thus
correlates with motility. In contrast, no direct correlation
appears to exist between the extent of actin polymerization,
locomotion and chemotaxis. The increase of cytoskeletal actin
induced by peptide at longer times of stimulation is lower
than that induced by phorbol ester, yet only the first agonist
induces front-tail polarity and efficient locomotion. PMA
induces only motility but not locomotion. A small net increase
in F-actin may be necessary for locomotion, but a specific
polar location of F-actin, for instance in the leading
lamellae, may be more important. One could envisage a dynamic
system involving shifts of F-actin from one cell area to
another.

SUMMARY

Different agonists induce motility and shape changes, but
only a specific polarized shape is correlated with directed
migration. An intact and dynamic actin network appears to be
important for motility and migration. Motility is usually
associated with an increased level of F-actin, and a specific
location of F-actin into surface protrusions. For locomotion,
a specific location of F-actin, rather than a large net
increase in F-actin appears to be of importance.

Three major groups of responses can be distinguished on
the basis of the type of shape changes, functional activity
and organization of F-actin.

1. Agents capable of polarizing cells, such as chemotactic
peptides, and microtubule-disassembling agents elicit, at
appropriate concentrations, a marked chemokinetic response,
but little if any fluid pinocytosis. F-actin shows a polar
location, being concentrated mainly in the protrusions at the
leading front. Chemotactic peptide also induces an increase
in the level of F-actin and cytoskeleton-associated actin. It
is, however, not clear if front-tail polarity and locomotion,
induced by chemotactic peptide after longer time of stimulation, correlate with an actual increase in the level of cytoskeleton-associated actin.

2. Activators of protein kinase C such as PMA and diacylglyce-

rols, induce nonpolar cells with surface projections. PMA and
diacylglycerols stimulate pinocytosis substantially. All three
agents tend to inhibit locomotion or chemotaxis as an immediate response. They also increase the percentage of
cytoskeletal actin, and induce an enrichment of F-actin in
surface projections.

3. Circus movement may occur in response to D2O. These cells
show little or no stimulation of locomotion or pinocytosis.
Thus the functional significance of this motor response
remains to be elucidated.

We conclude that different agonists can induce motility
and shape changes, but not necessarily chemotaxis. Only a polarized shape is correlated with directed locomotion. An
intact and dynamic actin network appears to be important for
motility including locomotion. Motility is usually associated
with an increased level of F-actin, and a specific location of
F-actin into surface protrusions. The actin-associated proteins α-Actinin, myosin and actin-binding protein appear also
to be important for pseudopod formation. For locomotion, a
specific location of F-actin, rather than a large net increase
in F-actin may be of importance.

ACKNOWLEDGMENT

The work was supported by the Swiss National Science Foundation. We thank Miss P. Kirschner, Miss G. Zürcher, Mr. D.
Meier, Miss M. Kilchenmann and Miss I. Lehmann for technical
assistance.

LITERATURE

Anderson, D. C., Wible, L. J., Hughes, B. J., Smith, C. W.,
Brinkley, B. R.: Cytoplasmic microtubules in polymorphonuclear leukocytes: Effects of chemotactic stimulation and
colchicine. <u>Cell</u> 31:719 (1982).

Boxer, L. A., Yoder, M., Bonsib, S., Schmidt, M., Ho, P.,
Jersild, R., Baehner, R. L.: Effects of a chemotactic factor,
N-formyl-methionyl peptide, on adherence, superoxide anion
generation, phagocytosis, and microtubule assembly of human
polymorphonuclear leukocytes. <u>J. Lab. Clin.</u> Med. 83:506
(1979).

Cox, C. C., Dougherty, R. W., Ganong, B. R., Bell, R. M.,
Niedel, J. E., Snyderman, R.: Differential stimulation of the
respiratory burst and lysosomal enzyme secretion in human
polymorphonuclear leukocytes by synthetic diacylglycerols.
<u>J. Immunol.</u> 136:4611 (1986).

Daukas, G., Lauffenburger, D. A., Zigmond, S.: Reversible
pinocytosis in polymorphonuclear leukocytes. <u>J. Cell Biol.</u>
96:1642 (1983).

Davis, B. H., Walter, R. J., Pearson, C. B., Becker, E. L.,
Oliver, J. M.: Membrane activity and topography of
f-Met-Leu-Phe-treated polymorphonuclear leukocytes. Acute and

sustained responses to chemotactic peptide. <u>Am. J. Pathol.</u> 108:206 (1982).

Davis, B. H., McCabe, E., Langweiler, M.: Characterization of f-Met-Leu-Phe-stimulated fluid pinocytosis in human polymorphonuclear leukocytes by flow cytometry. <u>Cytometry</u> 7:251 (1986).

Devreotes, P. N., Zigmond, S.H.: Chemotaxis in eukaryotic cells: a focus on leucocytes and dictyostelium. <u>Ann. Rev. Cell Biol.</u> 4: 649 (1988).

Euteneuer, U., Schliwa, M.: Evidence for an involvement of actin in the positioning and motility of centrosomes. <u>J. Cell Biol.</u> 101: 96 (1985).

Evans, J. F., Leblanc, Y., Fitzsimmons, B. J., Charleson, S., Nathaniel, D., Léveillé, C.: Activation of leukocyte movement and displacement of [3H]leukotriene B4 from leukocyte membrane preparations by (12R)-and 12S)-hydroxyeicosatetraenoic acid. <u>Biochim. Biophys. Acta</u> 917:406 (1987).

Fechtheimer, M., Zigmond, S. H.: Changes in cytoskeletal proteins of polymorphonuclear leucocytes induced by chemotactic peptides. <u>Cell Motil.</u> 3:349 (1983).

Fletcher, M. P., Seligmann, B. E., Gallin, J. I.: Correlation of human neutrophil secretion, chemoattractant receptor mobilization, and enhanced functional capacity. <u>J. Immun.</u> 128:941 (1982).

Ford-Hutchinson, A. W., Bray, M. A., Doig, M. V., Shipley. M. E., Smith, M. J.H.: Leukotriene B, a potent chemokinetic and aggregating substance released from polymorphonuclear leukocytes. <u>Nature</u> 286:264 (1980).

Gallin, J. I., Wright, D. G., Schiffman, E.: Role of secretory events in modulating human neutrophil chemotaxis. <u>J. Clin. Invest.</u> 62:1364 (1978).

Hafstrom, I., Palmblad, J., Malmsten, C. L., Radmark, O., Samuelsson, B.: Leukotriene B4 - A stereospecific stimulator for release of lysosomal enzymes from neutrophils. <u>FEBS Lett.</u> 130:146 (1981).

Haston, W. S., Shields, J.M.: Neutrophil leucocyte chemotaxis: a simplified assay for measuring polarising responses to chemotactic factors. <u>J. Immunol. Methods</u> 81:229 (1985).

Hoult, J. R. S., Nourshargh, S.: Phorbol myristate acetate enhances human polymorphonuclear neutrophil release of granular enzymes but inhibits chemokinesis. <u>Br. J. Pharmac.</u> 86:533 (1985).

Howard, T. H., Oresajo, C. O.: A method for quantifying F-actin in chemotactic peptide activated neutrophils: study of the effect of tBOC peptide. <u>Cell Motil.</u> 5:545 (1985).

Jesaitis, A. J., Bokoch, G. M., Tolley, J.O., Allen, R.A.: Lateral segregation of neutrophil chemotactic receptors into actin- and fodrin-rich plasma membrane microdomains depleted in guanyl nucleotide regulatory proteins. <u>J. Cell Biol.</u> 107:921 (1988).

Keller, H. U., Bessis, M.: Migration and chemotaxis of anucleate cytoplasmic leukocyte fragments. <u>Nature</u> 258:723 (1975).

Keller, H. U., Wilkinson, P. C., Abercrombie, M., Beckers, E. L., Hirsch, J. G., Miller, M. E., Ramsey, W. S., Zigmond, S. H.: A proposal for the definition of terms related to locomotion of leucocytes and other cells. <u>Clin. exp. Immunol.</u> 27:377 (1977).

Keller, H.U.: Motility, cell shape, and locomotion of neutrophil granulocytes. <u>Cell Motility</u> 3:47 (1983).

Keller, H. U., Zimmermann, A., Cottier, H.: Crawling-like movements, adhesion to solid substrata and chemokinesis of neutrophil granulocytes. <u>J. Cell Sci.</u> 64:8 (1983).

Keller, H.U., Naef, A., Zimmermann, A.: Effects of colchicine, vinblastine and nocodazole on polarity, motility, chemotaxis and cAMP levels of human polymorphonuclear leukocytes. <u>Exp. Cell. Res.</u> 153:173 (1984).

Keller, H. U., Zimmermann, A.: Shape, movement and function of neutrophil granulocytes. <u>Biomed. & Pharmacother.</u> 41:285 (1987).

Keller, H.U., Niggli, V., Zimmermann, A.: Diacylglycerols and PMA induce actin polymerization and distinct shape chang es in lymphocytes: relation to fluid pinocytosis and locomotion. <u>J. Cell Sci.</u> 93:457 (1989).

Keller, H. U., Niggli, V., Zimmermann, A., Portmann R.: The protein kinase C inhibitor H-7 activates human neutrophils: effect on shape, actin polymerization, fluid pinocytosis and locomotion. Subm. (1989).

Lehmeyer, J. E., Snyderman, R., Johnston, R. B.: Stimulation of neutrophil oxidative metabolism by chemotactic peptides: Influence of calcium ionconcentration and cytochalasin B and comparison with stimulation by phorbol myristate acetate. <u>Blood</u> 54:35 (1979).

Lewis, W.H.: On the locomotion of the polymorphonuclear neutrophils of the rat in autoplasma cultures. Bull. Johns <u>Hopkins Hosp.</u> 55:273 (1934).

Malawista, S. E.: Microtubule function in human blood polymorphonuclear leukocytes: analysis through heat-induced lesions. In: Dynamic Aspects of Microtubule Biology (Soifer D, ed), <u>Ann. NY Acad. Sci.,</u> Vol. 466, pp. 859 (1986).

Niggli, V., Jenni, V.: Actin-associated proteins in human neutrophils: identification and reorganization upon cell activation. <u>Eur. J. Cell Biol.</u> 49:366 (1989).

Nourshargh, S., Hoult, J. R. S.: Divergent effects of co-carcinogenic phorbol esters and a synthetic diacylglycerol on human neutrophil chemokinesis and granular enzyme secretion. <u>Br. J. Pharmac.</u> 91:557 (1987).

Omann, G. M., Allen, R. A., Bokoch, G. M., Painter, R. G., Traynor, A. E., Sklar, L.: Signal transduction and cytoskeletal activation in the neutrophil. Physiol. Rev. 67:285 (1987).

Parysek, C. M., Eckert, B. S.: Vimentin filaments in spreading, randomly locomoting and f-met-leu-phe-treated neutrophils. Cell Tissue Res. 235:575 (1984).

Rao, K. M., Varani, J.: Actin polymerization induced by chemotactic peptide and concanavalin A in rat neutrophils. J. Immunol. 129:1605 (1982).

Robinson, J. M., Badwey, J. A., Karnovsky, M. L., Karnovsky, M. J.: Cell surface dynamics of neutrophils stimulated with phorbol esters or retinoids. J. Cell Biol. 105:417 (1987).

Rollins, T. E., Zanolari, B., Springer, M. S., Guindon, Y., Zamboni, R., Lau, C.-K., Rokach, J.: Synthetic leukotriene B4 is a potent chemotaxin but a weak secretagogue for human PMN. Prostaglandins 25:281 (1983).

Roos, F. J., Zimmermann, A., Keller, H. U.: Effect of phorbol myristate acetate and the chemotactic peptide fNLPNTL on shape and movement of human neutrophils. J. Cell Sci. 88:399 (1987).

Schliwa, M., Pryzwanski, K. B., Euteneuer, U.: Centrosome splitting in neutrophils: an unusual phenomenon related to cell activation and motility. Cell 31:705 (1982).

Schliwa, M., Pryzwanski, K. B., Borisy, G. G.: Tumor promoter-induced centrosome splitting in human polymorphonuclear leukocytes. Eur. J. Cell Biol. 32:75-85 (1983).

Sha'afi, R. I., White, J. R., Molski, T. F. P., Shefzyk, J., Volpi, M., Naccache, P. H., Feinstein, M. B.: Phorbol 12-myristate 13-acetate activates rabbit neutrophils without an apparent rise in the level of intracellular free calcium. Biochem. Biophys. Res. Commun. 114:638 (1983).

Sha'afi, R. I., Molski, T. F. P.: Signalling for increased cytoskeletal actin in neutrophils. Biochem. Biophys. Res. Commun. 145:934 (1987).

Sheterline, P., Rickard, J. E., Boothroyd, B., Richards, R. C.: Phorbol ester induces rapid actin assembly in neutrophil leucocytes independently of changes in [Ca2+] and pHi. J. Muscle Res. Cell Motil. 7:405 (1986).

Showell, H. J., Freer, R. J., Zigmond, S. H., Schiffmann, E., Aswanikumar, S., Corcoran, B., Becker, E. L.: The structure-activity relations of synthetic peptides as chemotactic factors and inducers of lysosomal enzyme secretion for neutrophils. J. Exp. Med. 143:1154 (1976).

Spiegel, J. E., Schultz Beardsley, D., Southwick, F. S., Cux, S. E.: An analogue of the erythroid membrane skeletal protein 4.1 in non-erythroid cells. J. Cell Biol. 99:886 (1984).

Valerius, N. H., Stendahl, O., Hartwig, J. H., Stossel, T. P.: Distribution of actin-binding protein and myosin in polymor-phonuclear leucocytes during locomotion and phagocytosis. <u>Cell</u> 24:195 (1981).

White, J. G., Estensen, R.D.: Selective labilization of specific granules in polymorphonuclear leukocytes by phorbol myristate acetate. <u>Am. J. Pathol.</u> 75:45 (1974).

White, J. R., Naccache, P. H., Sha'afi, R. I.: The synthetic chemotactic peptide formyl-methionyl-leucyl-phenyl-alanine causes an increase in actin associated with the cytoskeleton in neutrophils. <u>Biochem. Biophys. Res. Commun.</u> 108:1144 (1982).

White, J. R., Huang, C.-K., Hill, J. Jr., Naccache, P. H., Becker, E. L., Sha'afi, R. I.: Effect of phorbol 12-myristate 13-acetate and its analogue 4α-phorbol 12,13-didecanoate on protein phosphorylation and lysosomal enzyme release in rabbit neutrophils. <u>J. Biol. Chem.</u> 259:8605 (1984).

Wright, T. M., Hoffman, R. D., Nishijima, J., Jakoi, L., Snyderman, R., Shin H. S.: Leukocyte chemoattraction by 1,2-diacylglycerol. <u>Proc. Natl. Acad. Sci.</u> 85:1869 (1988).

Zigmond, S. H., Sullivan, S. J.: Sensory adaptation of leukocytes to chemotactic peptides. <u>J. Cell. Biol.</u> 82:517 (1979).

Zigmond, S. H., Levitsky, H. I., Kreel, B. J.: Cell polarity: an examination of its behavioral expression and its consequen-ces for polymorphonuclear leukocyte chemotaxis. <u>J. Cell Biol.</u> 89:585 (1981).

Zimmermann, A., Keller, H. U., Cottier, H.: Heavy water (D2O)-induced shape changes, movements and F-actin redistribu-tion in human neutrophil granulocytes. <u>Eur. J. Cell Biol.</u> 47:320 (1988).

Zimmermann, A., Gehr, P., Keller, H. U.: Diacylglycerol-in-duced shape changes, movements and altered F-actin distribu-tion in human neutrophils. <u>J. Cell Sci.</u> 90:657 (1988).

LEUKODIAPEDESIS, COMPARTMENTALISATION AND SECRETION OF PMN LEUKOCYTE PROTEINASES, AND ACTIVATION OF PMN LEUKOCYTE PROCOLLAGENASE

H. Tschesche, B. Bakowski, A. Schettler,
V. Knäuper, H. Reinke and S. Krämer

Universität Bielefeld, Fakultät für Chemie
D-4800 Bielefeld 1, F.R.G.

INTRODUCTION

Human polymorphonuclear leukocytes (PMNL) are the first phagocytic cells to arrive at the site of injury of the epithelium, where they regulate the defense reactions against invading microorganisms. Therefore, their accumulation characterises an inflammatory process. Some particular properties of PMNL (Fig. 1) such as phagocytising microorganisms and eliminating damaged tissue are described [1].

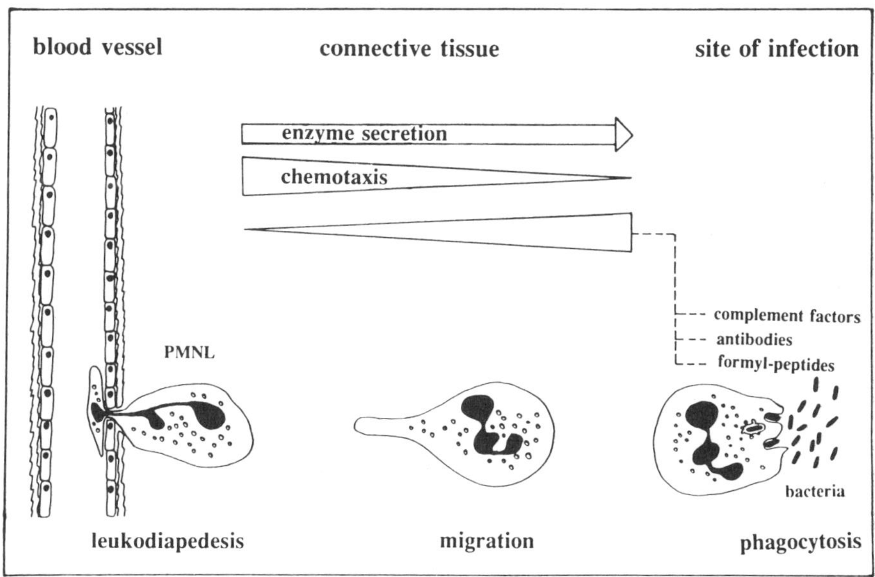

Fig. 1 Functions of PMNL

The induction of PMNL stimulation by binding of agonists, e. g. chemotactic N-formylpeptides, complement fragment C5a and antigen-antibody-complexes amongst others [2], to specific receptors on the cell surface initiates a signal transduction which leads to a microscopically visible change in the morphology of the cells. This polarisation is accompanied by a reorganisation of the microfilament system and a release of several enzymes and proteins from secretory vesicles [3, 4].

In response to chemotactic stimuli, PMNL of the marginated blood pool show an increased adherence to the endothelium of blood vessels, mediated by a leukocyte membrane glycoprotein, the Cdw 18-complex [5], and special adherence proteins. PMNL leave blood vessels through intercellular junctions of the endothelial cell layer [6]. After penetration of the underlying basement membrane the exudate PMNL migrate through the connective tissue to the site of infection. This process of exudation (leukodiapedesis) includes chemotactically induced locomotion (chemotaxis), which allows the migration of PMNL along a chemotactic gradient to the site of injury [7], where invading microorganisms and damaged tissue are eliminated by two processes.

One process for killing microorganisms is the "respiratory burst" [8], whereby cytotoxic superoxide and other oxide metabolites are generated. These also cause tissue damage and inflammation. The other is the more important function of PMNL, which marks them as phagocytes: Pathogenic particles are eliminated by engulfing them (phagocytosis) [9]. For intracellular digestion PMNL are provided with several proteolytic enzymes stored in different compartments - specific and azurophilic granules and the C-particles [10]. The most important enzymes in this context are elastase, collagenase and gelatinase [11-14], which are capable of degrading extracellular matrix components, such as type IV collagen, fibronectin amongst others.

Our investigations concern the unknown mechanism of the penetration of basement membranes and the release and storage of involved enzymes. In addition, we focus our attention on the activation mechanism of the secreted procollagenase, its purification, characterisation and protein structure.

The understanding of these unresolved problems is vital for medical research in the treatment of inflammatory diseases, such as rheumatoid arthritis.

LEUKODIAPEDESIS

To elucidate the mechanism involved in the penetration of basement membranes we used a model based on a modified Boyden chamber, of which two parts are divided by a micropore filter (pore size 5 μm) with an overlying human amnion membrane [15]. As this three-layered membrane is composed of an epithelial cell layer, a basement membrane and a loose stroma tissue it is suitable for studying the extravasation pathway of PMNL. By filling the lower compartment with FMLP (10^-7 M, formyl-methionyl-leucyl-phenylalanine), chemotactic migration of PMNL (in the upper compartment) through the membrane is induced, representing a model of the in vivo situation.

The investigation of this model by scanning electron microscopic studies gives a three-dimensional impression of the process of leukodiapedesis, especially the penetration of basement membranes after removal of the epithelial cell layer [16].

Figure 2 demonstrates a polarised PMNL with a rendered tail. This change in shape from an unstimulated cell with a smooth spherical form (not shown) is induced by the chemoattractant experienced from the cell via the gradient developed through the membrane.

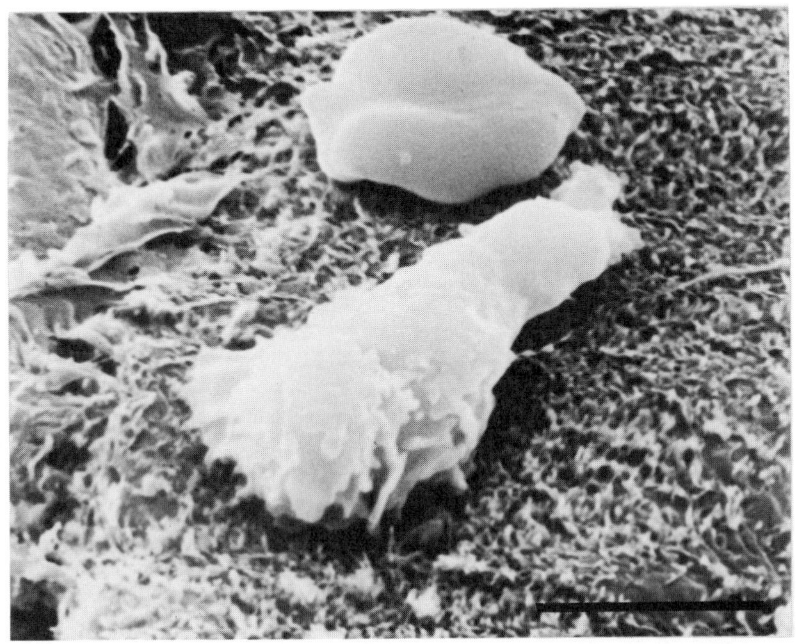

<u>Fig. 2</u>   Polarised PMNL on human amnion membrane
bar: 5 μm

The following two pictures (Fig. 3 and 4) illustrate the next step in the extravasation process. Tissue will be damaged in the immediate vicinity of the PMNL. The network of the basement membrane, consisting mainly of type IV-collagen, loses fibre density. This limited degradation prevents extensive, unnecessary destruction of the extracellular matrix.

Investigating the supernatants of FMLP-stimulated PMNL, we found increased amounts of gelatinase, which is the first enzyme to be secreted and detected in the extracellular environment [4]. Our findings confirm the results of Hibbs et al. [17] who assumed a sequential release of collagenolytic enzymes. The destruction of basement membranes is mainly caused by gelatinase, as was shown by the in vitro investigations of Vissers et al. [18], Uitto et al. [19] and Tschesche et al. [20], and which concurs with our results with this semi-physiological system [16]. Wright and Gallin [21] already demonstrated that migration of PMNL is accompanied by exocytosis of specific granules.

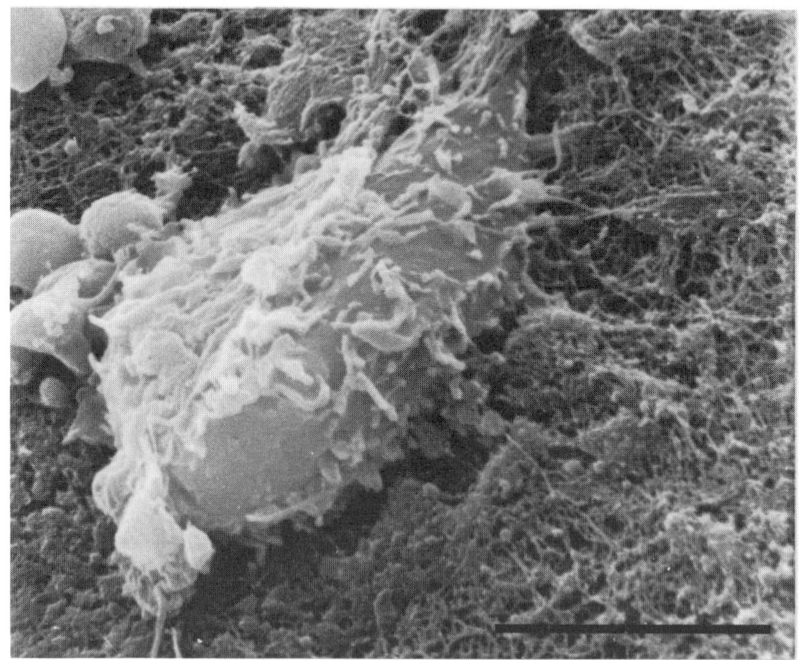

Fig. 3 Polarised PMNL, starting to lyse tissue
        bar: 5 μm

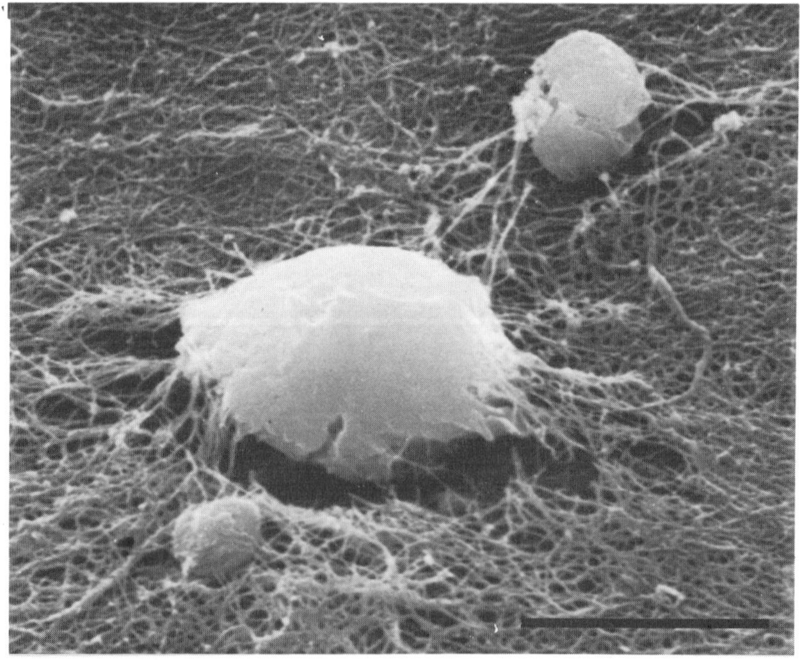

Fig. 4 Damage of tissue is limited to the immediate
        environment of the cell
        bar: 5 μm

Another strong indication that metalloproteinases are
involved in membrane penetration is supplied by inhibition
experiments [16]. Preincubation of the membrane with TIMP
(tissue inhibitor of metalloproteinases) leads to an inhibited migration of PMNL through the amnion membrane.

Partial degradation of the matrix seems to be a prerequi-
site for PMNL penetration and facilitates their locomotion.
Besides limited proteolytic degradation of the basement
membrane barrier, PMNL may use mechanical forces for penetration. The present picture of a PMNL locomoting through the
membrane (Fig. 5) indicates both partial destruction and
active mechanical dilatation of the matrix network.
Observations of PMNL locomoting through the loose stroma
tissue (not shown) indicated little or no destruction of this
type I-collagen network. These stroma fibres are obviously no
obstacle for migration to the site of infection.

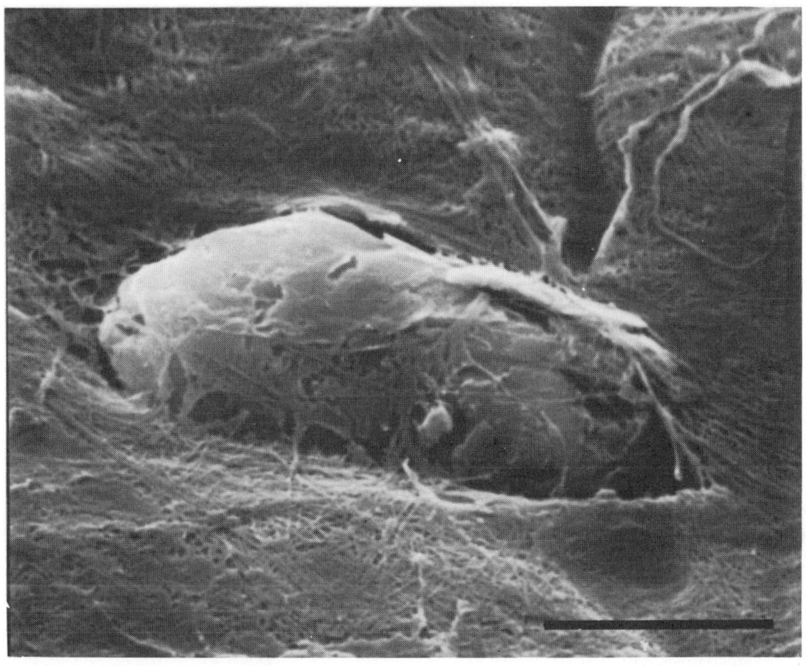

<u>Fig. 5</u> Involvement of enzymatic degradation and active
         mechanical dilatation of the fibre matrix
         bar: 5 μm

Our observations lead to the conclusion, that PMNL
probably use the same process as tumour cells while leaving
blood vessels. Liotta et al. [22] postulated a three-step-
process for tumour invasion. Firstly, tumour cells adhere at
the endothelial cells of the membrane, slip through between
the cells and then locally degrade matrix by proteolytic
enzymes, the type IV-collagenase/gelatinase. Penetration of
the basement membrane could be inhibited by TIMP, as we could
demonstrate for PMNL. Finally, the cells migrate through the
connective tissue.

RELEASE AND INTRACELLULAR COMPARTMENTALISATION OF PROTEINASES

Stimulation of PMNL induces a reorganisation of their microfilament system and leads to a release of several enzymes and proteins. The involvement of the cytoskeleton in this process is the subject of several investigations and publications (for review [23]) but a complete resolution has not yet been found. In addition, there are contradictions with regard to the intracellular compartmentalisation of the proteinases involved. Most authors distinguish between azurophilic or primary and specific or secondary granules. The former contain elastase, myeloperoxidase and several hydrolytic enzymes, the latter collagenase, lactoferrin, vitamin-B_{12}-binding protein and lysozyme [24]. A third type of granule has been described - C-particles with gelatinase as a marker enzyme [25]. The storage of these enzymes in either one of these three types of granule has not yet been clarified beyond doubt.

Therefore, two sets of experiments were performed to clarify these questions on the association of different granules with the cytoskeleton and on the sucessive release of the granula proteins.

Density gradient centrifugation was used to separate sedimentable polymerised microfilaments and microtubules. Determinations of various marker enzymes in the sediment and supernatant revealed that proteins were distributed unequally, as seen in Fig. 6. Elastase, myeloperoxidase and lactoferrin were located in the cytoskeleton fraction and, therefore, indicated the association of these granules with the cytoskeleton. The supernatant contained most of the metalloproteinases which provides evidence that these enzymes are stored in another type of compartment.

Disruption of the cytoskelton with cytochalasin B (CB) without additional stimulation with FNLPNTL (10^{-7} M, formyl-norleucyl-leucyl-phenylalanyl-norleucyl-tyrosyl-leucine) led to no significant increase in protein release into the supernatant, except in the case of gelatinase, where approximately 12 to 20 % of total gelatinase content was secreted. Treatment with FNLPNTL alone yielded similar results. However, a dramatic change in protein release was observed when PMNL were preincubated with CB and subsequently stimulated with FNLPNTL. The supernatant of these cells contained increased amounts of all proteins under examination. These findings are summarised in Fig. 7. This diagram shows a remarkable difference between the release of elastase and myeloperoxidase, which is secreted almost completely, whilst only about 25 % of total elastase content can be measured in the supernatant. These data could lead to the conclusion that the two enzymes are not stored and released from the same granule type, if the phenomenon is not due to specific adsorption of elastase, e. g. on the membrane surface, which has to be clarified.

Furthermore, we studied the secretion of proteinases and lactoferrin, which was induced by phagocytosis and/or FNLPNTL, for 60 minutes. For this purpose, cells were preincubated with opsonised zymosan particles. The secretion profiles are shown in Fig. 8.

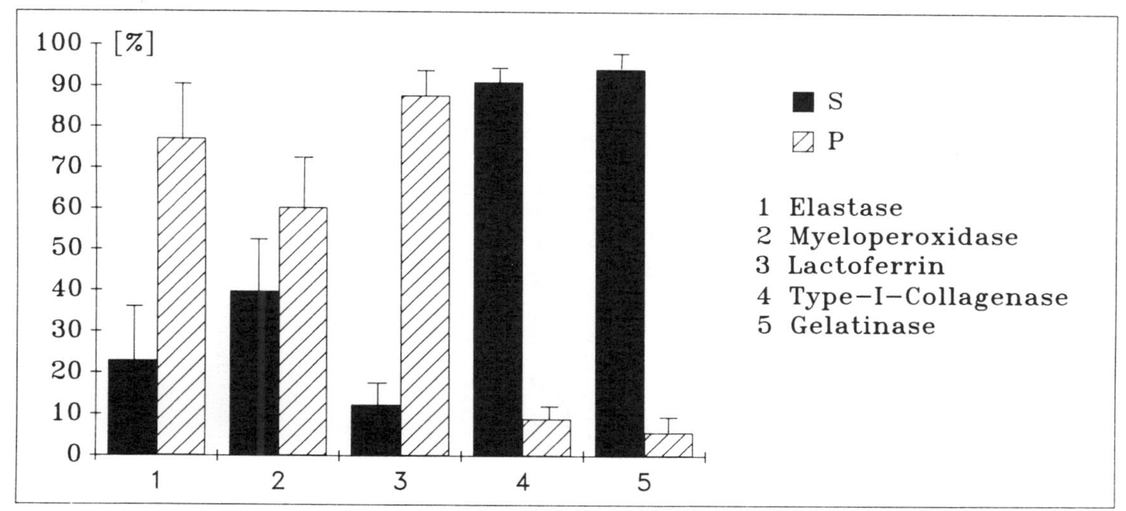

Fig. 6   Enzyme distribution after differential centrifugation. Cells were preincubated for 20 min at 37 ^{o}C before low energy sonication. Free granules and granules associated with the cytoskeleton were separated by centrifugation.
Enzymes were released by high energy sonication and detergent treatment.
Each value represents the mean of eleven experiments (eight to nine different donors).
S, supernatant   P, pellet

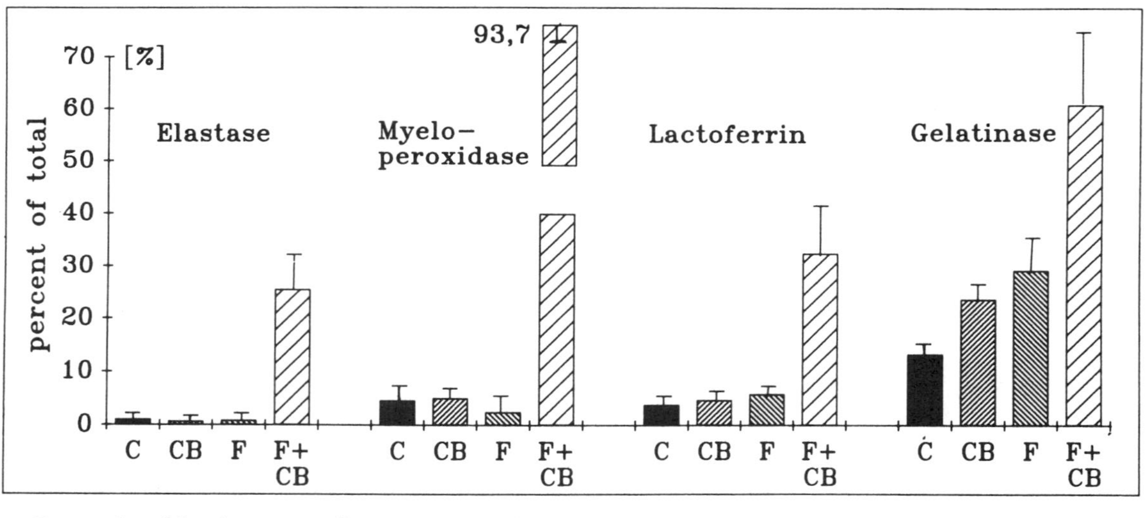

Fig. 7 Granule discharge of stimulated PMNL after preincubation with cytocholasin B (CB). After incubation (20 min/37 $^{\circ}$C) in suspension without or with CB (5 µg/ml) the PMNL were plated into culture dishes supplemented with cytochalasin B (CB), FNLPNTL (10^{-7} M) (F) or both agents (F+CB) for 10 min; C, untreated controls.
The data shown were obtained in eleven different experiments (seven donors).

In contrast to elastase and lactoferrin, of which only
6 % of the total amounts were released, approximately 40 % of
total collagenase could be detected in the supernatant of
stimulated PMNL.

These results give further evidence for different
compartmentalisation of enzymes involved in matrix degradation
(elastase, gelatinase and collagenase) and those participating
in the burst reaction (myeloperoxidase and lactoferrin).

Our results [26] confirm previous suggestions on the
existence of granule subclasses based on immunoelectron
microscopic [27] and biochemical [28] examinations. In
addition, centrifugation in our laboratory of PMNL cytosolic
content in a continous density gradient revealed separation
into seven distinct granula populations. Marker enzymes, such
as elastase and collagenase are localised in more than one
fraction of the distinct granule populations [29].

CHARACTERISATION AND ACTIVATION OF PROCOLLAGENASE

Several inflammatory diseases are accompanied by tissue
damage mainly caused by PMNL proteolytic enzymes. As mentioned
above, PMNL store proteinases in distinct granules, whose
secretions are responsible for the breakdown of the extracellular matrix. As a component of the specific granules,
collagenase especially cleaves the type I, II and III collagen
network of connective tissue. Thus, we focused our attention
on the isolation and characterisation of this enzyme [30];
particularly, as there is still conjecture as to the molecular
mass, physicochemical, and inhibitory properties and sequence
information. In addition, the activation mechanism is still
under investigation.

A PMNL procollagenase was purified by a rapid and repro-
ducible method including affinity chromatography on zinc
chelate Sepharose, and ion exchange chromatography on Q-
Sepharose fast flow, followed by affinity chromatography on
orange Sepharose and a gel-permeation step on Sephacryl S-300
[30].

The proenzyme was isolated in a latent form and consisted
of a single polypeptide chain with M_r 85 000, as shown by
SDS/PAGE under reducing and non-reducing conditions.
Digestion with endoglycosidase F resulted in a decrease in the
apparent molecular mass to a value of 53 000, which is
approximately identical to the M_r of the unglycosylated form
of the fibroblast collagenase [31].

The complete primary structure of the propeptide region
(Fig. 9) exhibits a 47 % homology to the synovial cell [32]
and fibroblast collagenase [33]. The overall homology of the
partial amino acid sequence data of PMNL procollagenase and
fragments so far elucidated was estimated to be 55 % [34],
indicating an ancestral relationship but also independent
genes for both enzymes.

Information on the proteolytic activation of PMNL procollagenase was obtained from N-terminal sequence determinations

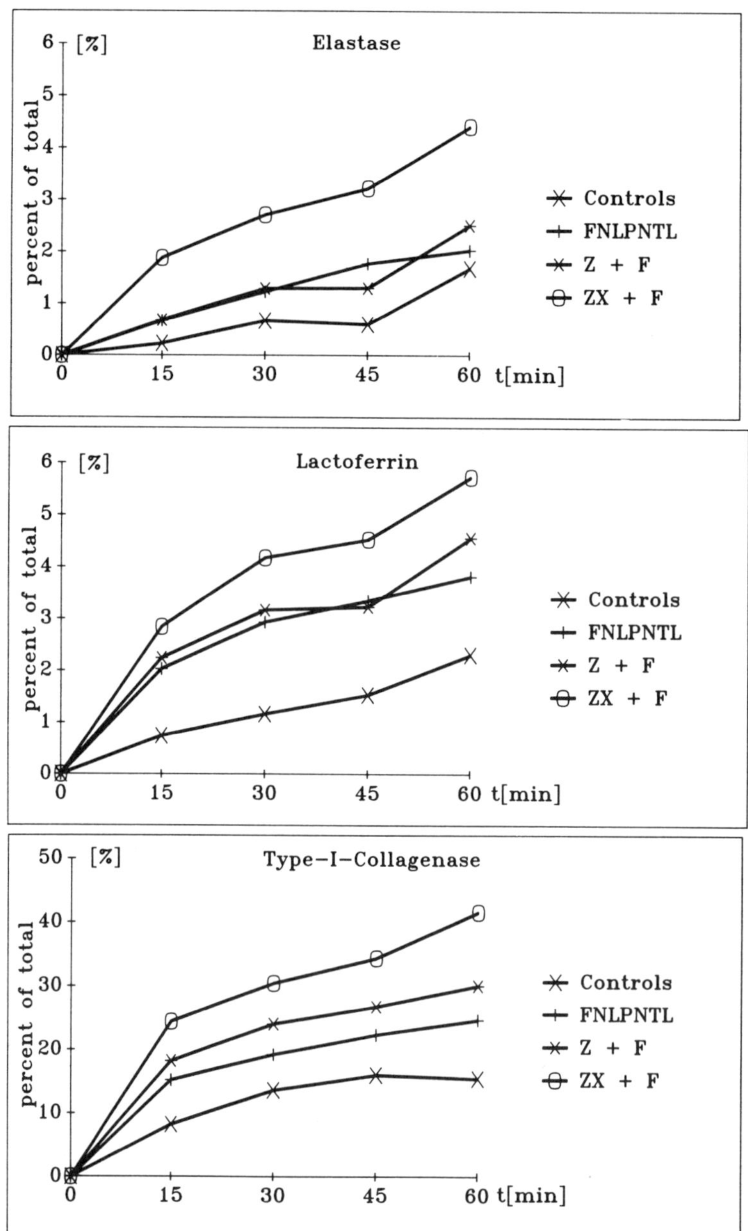

```
Fibroblast   F P A T L E T Q E Q D V D L V Q K Y L E K Y Y N L K N D G R Q V E K R R N S G P V V E K L K Q M Q E

             1                 10                  20                  30                  40
Leucocyte    F P - - V S S K E K N T K I V Q D Y L E K I Y Q L P S N Q V Q S T R K I G T N V I V E K L K E M Q R↑
                                                                                                 Trypsin
                                                                                                 (Mr 70000)

Fibroblast   F F G L K V T G K P D A E T L K V M K Q P R C G V P D V A Q F V L T E G N P R W E Q T H L T Y R I E

               50                60                  70                  80                  90
Leucocyte    F F G L G↑V T G K P N E E T L D M M E/K K P R↑C G V P D S G G F↑M L↑T P G N P K W E R T ? L T Y R I R
                     Elastase                        Trypsin           |    Kallikrein               -------
                     (Mr 66000)                      (Mr 65000)        |    (Mr 64000)
                                                                 Chymotrypsin
                                                                 Cathepsin G
                                                                 (Mr 64000)

Fibroblast   N Y T P D L P R A D V D H A I E K A F Q L W S N V T P L T F T K V S E G Q A D I M I S F V R G D H R
             -----                                         -----
               100               110                 120
Leucocyte    N Y T P Q L S E A E V E R A I K D A F E L W
             -----
```

Fig. 9 N-terminal sequence comparison of fibroblast and PMNL collagenase. Activation sites are indicated by arrows. Possible carbohydrate attachment sites are underlined.

of the enzyme activated by trypsin, chymotrypsin, pancreatic
kallikrein, or PMNL cathepsin G, that all led to truncated
active forms of the PMNL procollagenase with reduced molecular
mass. Fig. 9 shows the proteolytic cleavage sites by these
serine proteinases. Activation by trypsin led, in the first
instance, to a still latent intermediate form with M_r 70 000
before being converted into the active enzyme with M_r 65 000.
We did not find evidence for involvement of an autoproteolytic
activation step, as assumed for the activation of fibroblast
collagenase [35]. However, activation by tissue kallikrein
resulted in a final cleavage of a Leu^{81} - Thr^{82} peptide bond,
which is not characteristic of the usual specifity of kallikrein. Therefore, it cannot be excluded that an autoproteolytic process takes place, similar to the mercurial activation
of procollagenase.

Our observations (published in detail [30]) suggest the
following hypothetical model.

Activation of PMNL collagenase can be generated by
proteolytic cleavage in a limited strongly conserved region
(PRCGVPD) around residues 70 - 82, which then leads to a
change in the coordination sphere of the integral metal ion
(Zn) opening the catalytic site and generating enzymatic
activity. Vallee and Auld [36] assumed a coordination of the
metal ion by three amino acid side chains and one "activated"
water molecule as the structural characteristic of the active
site of metalloproteinases.

The latent proenzyme contains the sequence Pro-Arg-Cys-
Gly-Val-Pro-Asp, which may be responsible for the correct
position of the propeptide, whereas both proline residues are
considered to maintain latency. The free Cys^{71} residue
possibly stabilises the tertiary structure of procollagenase
by coordinating the metal ion at the active site of the
enzyme.

Further investigations, especially crystallographic
determinations, must be carried out to support this hypothesis.

Although our results contribute to the understanding of
in vitro secretion and activation of enzymes involved in
matrix degradation, further investigations are needed to
elucidate the in vivo mechanisms of activation.

ACKNOWLEDGEMENTS

This work was supported by the Deutsche Forschungsgemeinschaft (SFB 223).
The authors wish to thank Mrs. G. Delany for linguistic
advice.

REFERENCES

1. Gallin, J. I., Goldstein, I. M., Snyderman, R.: Inflam-
    mation: Basic Principles and Clinical Correlates. Raven
    Press, New York (1988).

2. Schmitt, M., Keller, H. U., Cottier, H.: Qualitative and quantitative assessment of human polymorphonuclear (PMNL) functions. <u>Beitr. Infusionstherapie Klin. Ernähr.</u> 15:196-230 (1987).

3. Carson, M., Weber, A., Zigmond, S. H.: An actin-nucleating activity in polymorphonuclear leukocytes is modulated by chemotactic peptides. <u>J. Cell Biol.</u> 103:2707-2714 (1986).

4. Tschesche, H., Schettler, A., Thorn, H., Bakowski, B., Knäuper, V.; Reinke, H.; Jockusch, B. M.: Chemotaxis, proteinase secretion, and activation of collagenase of PMN leukocytes. In: E. Auerswald, H. Fritz, V. Turk (eds.): Bilateral Seminars of the International Bureau Kernforschungsanlage Jülich GmbH: Proceedings of the 8th Winterschool on "Proteinases and their Inhibitors". KFA-Jülich Verlag. Jülich. pp. 31-36 (1989).

5. Harlan, J. M., Schwartz, B. R., Wallis, W. J., Pohlman, T. H.: The role of neutrophil membrane proteins in neutrophil emigration. In: H. Z. Movat (ed.): Leukocyte Emigration and its Sequela. Satellite Symp. 6th Int. Congr. Immunology, Toronto, Ont. 1986. Karger, Basel. pp. 94-104. (1987).

6. Cramer, E. B., Milks, L. C., Ojakian, G. K.: Transepithelial migration of human neutrophils: an in vivo model system. <u>Proc. Natl. Acad. Sci. USA</u> 77:4069-4073 (1980).

7. Wilkinson, P. C., Haston, W. S.: Chemotaxis: An overview. In: G. Di Sabato (ed.): Methods in Enzymology. Immunochemical Techniques. Part L. Chemotaxis and Inflammation. <u>Academic Press</u>, San Diego, Vol. 162, pp. 3-16 (1988).

8. Baggiolini, M., Wymann, M. P.: Turning on the respiratory burst. <u>TIBS</u> 15:69-72 (1990).

9. Hoffstein, S. T.: Intra- and extracellular secretion from polymorphonuclear leukocytes. In: G. Weissm.ann (ed.): The Cell Biology of Inflammation. Elsevier/North-Holland Biomedical Press. Amsterdam, New York, Oxford, pp. 387-430 (1980).

10. Wright, D. G.: In: G. Di Sabato (ed.): Human neutrophil degranulation. Methods in Enzymology. Immunochemical Techniques. Part L. Chemotaxis and Inflammtion. Academic Press, San Diego, pp. 162:538-551 (1988).

11. Birkedal-Hansen, H. J.: From tadpole collagenase to a family of matrix metalloproteinases. <u>Oral Pathol.</u> 17:445-451 (1988).

12. Kohnert, U., Oberhoff, R., Fedrowitz, J., Bergmann, U., Rauterberg, J., Tschesche, H.: The degradation of collagen by a metalloproteinase from human leucocytes. In: W. H. Hörl, A. Heidland (eds.): Proteases II: Potential Role in Health and Disease, Advances in Experimental Medicine and Biology. Plenum Press, New York, pp. 240:33-44 (1988).

13. Tschesche, H., Knäuper, V., Krämer, S., Michaelis, J., Oberhoff, R., Reinke, H.: Latent collagenase and gelatinase from human neutrophils and their activation. In: H. Birkedal-Hansen, Z. Werb, H. Welgus, H. van Wart (eds.): Matrix Metalloproteinases and Inhibitors. Gustav Fischer Verlag. Stuttgart, New York, in press (1990).

14. Stein, R. L., Trainor, D. A., Wildonger, R. A.: Neutrophil elastase. <u>Ann. Rep. Med. Chem.</u> 20:237-246 (1985).

15. Russo, R. G., Liotta, L. A., Thorgeirsson, U., Brundage,

R., Schiffmann, E.: Polymorphonuclear leukocyte migration through human amnion membrane. J. Cell Biol. 91:459-467 (1981).

16. Bakowski, B.: Untersuchungen zur Migration menschlicher polymorphkerniger neutrophiler Granulocyten durch Basalmembranen. PhD Thesis. Universität Bielefeld. (1990).

17. Hibbs, M. S., Hasty, K. A., Seyer, J. M., Kang, A. H., Mainardi, C. L.: Biochemical and immunological characterization of the secreted forms of human neutrophil gelatinase. J. Biol. Chem. 260:2493-2500 (1985).

18. Vissers, M. C. M., Winterbourn, C. .C., Hunt, J. S.: Degradation of glomerular basement membrane by human neutrophils in vitro. Biochim. Biophys. Acta 804:154-160 (1984).

19. Uitto, V.-J., Schwartz, D., Veis, A.: Degradation of basement-membrane collagen by neutral proteases from human leukocytes. Eur. J. Biochem. 105:409-417 (1980).

20. Tschesche, H., Fedrowitz, J., Kohnert, U., Macartney, H. W., Michaelis, J., Kühn, K., Wiedemann, H.: Interstitial collagenase, gelatinase and a specific type IV/V (basement membrane) collagen degrading proteinase from human leukocytes. In: H. Tschesche (ed.): Proteinases in Inflammation and Tumor Invasion: Review Articles. pp. 225-243. Walter de Gruyter. Berlin, New York. (1986).

21. Wright, D. G., Gallin, J. I.: Secretory responses of human neutrophils: exocytosis of specific (secondary) granules by human neutrophils during adherence in vitro and during exudation in vivo. J. Immunol. 123:285-294 (1979).

22. Liotta, L. A., Guirguis, R., Stracke, M.: Review article: Biology of melanoma invasion and metastasis. Pigment Cell Res. 1:5-15 (1987).

23. Anderson, D. C., Wible, L. J., Hughes, B. J., Smith, C.W., Brinkley, B. R.: Cytoplasmic microtubules in polymorphonuclear leukocytes: effects of chemotactic stimulation and colchicine. Cell 31:719-729 (1982).

24. Henson, P. M., Henson, J. E., Fittschen, C., Kimani, G., Bratton, D. L., Riches, D. W. H.: Phagocytic cells: Degranulation and secretion. In: J. I. Gallin, I. M. Goldstein, R. Snyderman (eds.): Inflammation: Basic Principles and Clinical Correlates. Raven Press, New York, pp. 363-390 (1988).

25. Dewald, B., Bretz, U., Baggiolini, M.: Release of gelatinase from a novel secretory compartment of human neutrophils. J. Clin. Invest. 70:518-525 (1982).

26. Schettler, A., Thorn, H., Jockusch, B. M., Tschesche, H.: Release of proteinases from stimulated PMNL: Evidence for subclasses of the main granule types and their association with cytoskeletal components. Eur. J. Biochem., submitted (1990).

27. Hibbs, M. S., Bainton, D. F.: Human neutrophil gelatinase is a component of specific granules. J. Clin. Invest. 84:1395-1401 (1989).

28. Perez, D. H., Marder, S., Elfman, F., Ives, E. H.: Human neutrophils contain subpopulations of specific granules exhibiting different sensitivities to changes in cytosolic free calcium. Biochem. Biophys. Res. Commun. 145:976-981 (1987).

29. Nitsch, M., Gabrijelcic, D., Tschesche, H.: Separation of Granule Subpopulations in Human Polymorphonuclear

Leucocytes. <u>Biol. Chem. Hoppe-Seyler</u> 371:611-615 (1990).

30. Knäuper, V., Krämer, S., Reinke, H., Tschesche, H.: Characterization and activation of procollagenase from human polymorphonuclear leucocytes. N-terminal sequence determination of the proenzyme and various proteolytically activated forms. <u>Eur. J. Biochem.</u> 189:295-300 (1990).

31. Wilhelm, S. M., Eisen, A. Z., Teter, M., Clark, S. D., Kronberger, A., Goldberg, G. I.: Human fibroblast collagenase: glycosylation and tissue-specific levels of enzyme synthesis. <u>Proc. Natl. Acad. Sci. USA</u> 83:3756-3760 (1986).

32. Brinckerhoff, C. E., Ruby, P. L., Austin, S. D., Fini, M. E., White, H. D.: Molecular cloning of human synovial cell collagenase and selection of a single gene from genomic DNA. <u>J. Clin. Invest.</u> 79:542-546 (1987).

33. Goldberg, G. I., Wilhelm, S. M., Kronberger, A., Bauer, E. A., Grant, G. A., Eisen, A. Z.: Human fibroblast collagenase. Complete primary structure and homology to an oncogene transformation-induced rat protein. <u>J. Biol. Chem.</u> 261: 6600-6605 (1986).

34. Knäuper, V., Krämer, S., Reinke, H., Tschesche, H.: Partial amino acid sequence of human PMN leukocyte procollagenase. <u>Biol. Chem. Hoppe-Seyler (Suppl.)</u> 371:295-304 (1990).

35. Grant, G. A., Eisen, A. Z., Marmer, B. L., Roswit, W. T., Goldberg, G. I.: The activation of human skin fibroblast procollagenase. Sequence indentification of the major conversion products. <u>J. Biol. Chem.</u> 262:5886-5889 (1987).

36. Vallee, B. L., Auld, D. S.: Short and long spacer sequences and other structural features of zinc binding sites in zinc enzymes. <u>FEBS Lett.</u> 257:138-140 (1989).

ROLE OF PLATELET ACTIVATING FACTOR IN THE ADHESION PROCESS

OF POLYMORPHONUCLEAR NEUTROPHILS TO ENDOTHELIAL CELLS

Federico Bussolino, Daniela Alessi,
Ernesto Turello, Giovanni Camussi

Dipartimento di Genetica, Biologia e Chimica
Medica, Laboratorio di Immunopatologia,
Universita' di Torino; Dipartimento di Biochimica
e Biofisica, Universita' di Napoli

INTRODUCTION

Circulating polymorphonuclear neutrophils (PMN) provide a
front line of defense that can be rapidly mobilized and
activated against infectious and toxic agents. The first step
in extravasion involves the adhesion of neutrophils to vascular endothelium. This process must be regulated to allow
localization of neutrophils only to inflammatory sites.
Chemotactic factors, cytokines or lipid mediators released at
the inflamed sites may modify the characteristics of plasma-
membrane surface of PMN, endothelial cells (EC) or of both and
promote PMN-EC adhesion (reviewed in 1). For example, the
leukocyte CD11/CD18 surface adhesive complex and its ligand
(intercellular adhesion molecule 1) on EC surface, are essential in the PMN extravasion process and their expression can
be up-regulated by cytokines and chemotactic peptides (2-4).
Interleukin-1 and tumor necrosis factor induce a transient
expression of the glycoprotein endothelial-leukocyte adhesion
molecule-1 on EC surface that mediate PMN-EC interaction
(4,5).

Recent reports indicate that platelet activating factor
(PAF, 1-0-alkyl-2-acetyl-sn-glycero-3-phosphocholine), a mediator of inflammation and intercellular communication (reviewed in 6), increases the adhesion of leukocytes to endothelium (7,8) and to inert surfaces (9), and may, at least,
in part mediate the thrombin-induced PMN adhesion to EC (10).
Since both PMN and EC synthesize and act as a target for PAF
(6), PAF produced after appropriate stimulation by PMN and EC
may partially be involved in modulating PMN-EC interaction.

This study was undertaken to evaluate: 1) the efficiency
of stimuli for PAF synthesis in promoting the adhesion of PMN
to EC; 2) the primary cell target of PAF in the process of PMN
adhesion to EC; 3) the relative contribution and the specificity of PAF produced by the two cell types.

New Aspects of Human Polymorphonuclear Leukocytes
Edited by W.H. Hörl and P.J. Schollmeyer, Plenum Press, New York, 1991

MATERIALS AND METHODS

Cell preparation

Human PMN were prepared from blood by centrifugation as described (11). Pelleted cells were resuspended in 2 volumes of 2.5 % gelatin (Difco) in saline. The bulk of erythrocytes was removed by low speed centrifugation and subsequent osmotic shock. PMN (> 90 % pure) at 1×10^7/ml were resuspended in Hanks' balanced salt solution (HBSS), Ca/Mg-free, containing HEPES 20 mM, pH 7.4, and labeled for 30 min at room temperature with 10 μCi/ml ^{51}Cr (Amersham). After radiolabeling, the cells were washed twice with HBSS, Ca/Mg-free, containing 0.25 % bovine serum albumin (BSA, Sigma), and resuspended in the same buffer. Human EC from umbilical cord veins were grown and characterized as described (11) and used at I-II passage.

PMN-adhesion assay

Confluent EC (1 to 1.5×10^5 in a 2 cm^2 culture well) were washed twice with HBSS, Ca/Mg-free and incubated with 0.4 ml medium 199 (Gibco) containing 0.25 % BSA and 0.1 ml PMN suspension for 15 min at 37 ^{0}C. At the end of incubation the supernatant was carefully aspirated, the wells were washed twice with 1 ml of HBSS containing 0.25 % BSA to remove non adherent PMN, and incubated for at least 10 min with 0.25 ml of NaOH 1 N + 1 % sodium dodecyl sulphate and counted using a gamma counter (12).

Adhesion assay was carried out in the following experimental conditions: 1) stimulation of PMN co-incubated with EC with n-formyl-methionyl-leucyl-phenylalanine (FMLP), A23187, thrombin or angiotensin II (Sigma); 2) stimulation of PMN or EC followed by two washes to remove the agonist before starting the adhesion assay. In some experiments PMN or EC were preincubated for 15 min at 37 ^{0}C with two different PAF receptor antagonists, CV-3988 (Takeda) and BN52021 (Institut Henri Beaufour). Alternatively, these drugs were added for 15 min in co-stimulatory experiments before the addition of the agonist. For the desensitization study, PMN suspended in HBSS, Ca/Mg-free, containing 0.25 % BSA were incubated with 50 nM PAF (Bachem, 1-0-octadecyl-2-acetyl-sn-glycero-3-phosphocholine) for 10 min at 37 ^{0}C. After this period, PMN were washed with HBSS, Ca-/Mg-free and used for the adhesion assay.

PAF production

PAF released into the medium and cell-associated was isolated, characterized and measured as previously described (13).

RESULTS

Thrombin, FMLP, angiotensin II and A23187 enhanced PMN adhesion to EC co-stimulatory conditions. Thrombin, FMLP and angiotensin II caused a rapid increased in adhesion (Fig. 1) in a dose dependent manner (Table 1). The action of A23187 was slower (Fig. 1).

Table 1. Dose-response effect of thrombin, angiotensin II, and
         FMLP stimulation of PMN adherence to EC in co-
         stimulatory condition

Stimulus		PMN adherence to EC[a] (PMN bound/mm^2)
Control		323 ± 89
Thrombin	0.01 U/ml	490 ± 110
	0.1 U/ml	1256 ± 183
	0.5 U/ml	1677 ± 245
	1.0 U/ml	1934 ± 178
Angiotensin II	1 nM	398 ± 112
	10 nM	871 ± 111
	100 nM	1231 ± 210
FMLP	1 nM	489 ± 90
	10 nM	679 ± 123
	50 nM	1267 ± 202
	100 nM	1689 ± 165

[a]Mean ± S.D. of 4 experiments done in triplicate

Preincubation of EC with thrombin (0.5 U/ml for 3 min),
angiotensin II (100 nM for 10 min), A23187 (1 μM for 15 min),
but not with FMLP (50 nM for 10 min), enhanced PMN adhesion to
EC. The results obtained with thrombin and angiotensin II in
these experiments are similar to those obtained in co-stimulatory conditions. In contrast, A23187 shows a less marked activity than in co-stimulatory experiments (Table 2). Experiments
of preincubation were also performed with PMN. The pretreat-

Table 2.   PMN adhesion to EC

Condition[a]		PMN bound/mm^2
PMN STIMULATION		
None		341 ± 100
Thrombin	(0.5 U/ml, 5 min)	398 ± 109
Angiotensin II	(100 nM , 10 min)	298 ± 189
FMLP	(100 nM , 10 min)	1056 ± 271
A23187	(1 μM , 10 min)	1120 ± 167
EC STIMULATION		
None		451 ± 123
Thrombin	(0.5 U/ml, 5 min)	1737 ± 201
Angiotensin II	(100 nM , 15 min)	1109 ± 110
FMLP	(100 nM , 10 min)	321 ± 78
A23187	(1 μM , 30 min)	1043 ± 56

[a]PMN or EC were preincubated as indicated before the adhesion
  assay. Mean ± S. D. of 3 experiments done in triplicate.

ment of PMN with FMLP (50 nM) and A23187 (1 μM) for 10 min increased the base line adherence. However, the effects of these stimuli were less effective than that in co-stimulatory experiments. Preincubation of PMN with thrombin or angiotensin II was ineffective (Table 2).

Relationship between PAF production and PMN adhesion

The time-course of PAF production by EC and PMN stimulated with an optimal concentration of agonist giving the maximal adhesion of PMN is shown in Figures 2 and 3.

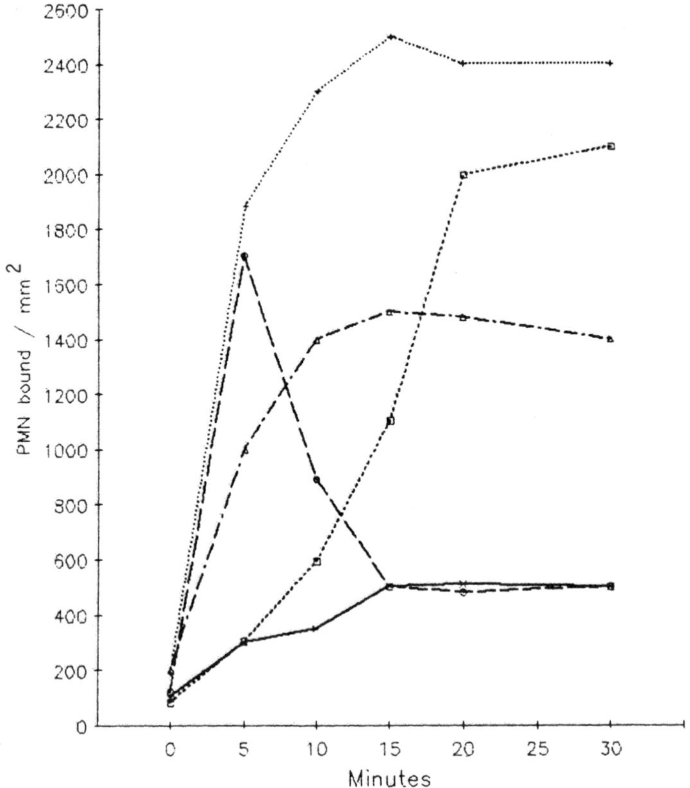

Fig. 1   Time course of PMN adhesion to EC in co-stimulatory conditions. Control **x** , thrombin (0.5 U/ml) ◯ , angiotensin II (100 nM) △ , FMLP (100 nM) **+** , A23187 (1 μM) □ . Data shown are representative of 5 experiments done with similar results.

Maximal production of PAF by EC occurred after 3–5 min, 15 min and 30 min stimulation with thrombin, angiotensin II and A23187, respectively. Thrombin activity declined to basal value after 10 minutes, whereas the effect of angiotensin II and A23187 was more protracted. FMLP (0.05–2 μM) did not activate PAF synthesis.

PMN produce PAF when stimulated with FMLP or A23187, but
not with thrombin or angiotensin II. The peak of PAF production is reached after 5 and 15 min incubation with FMLP and
A23187, respectively.

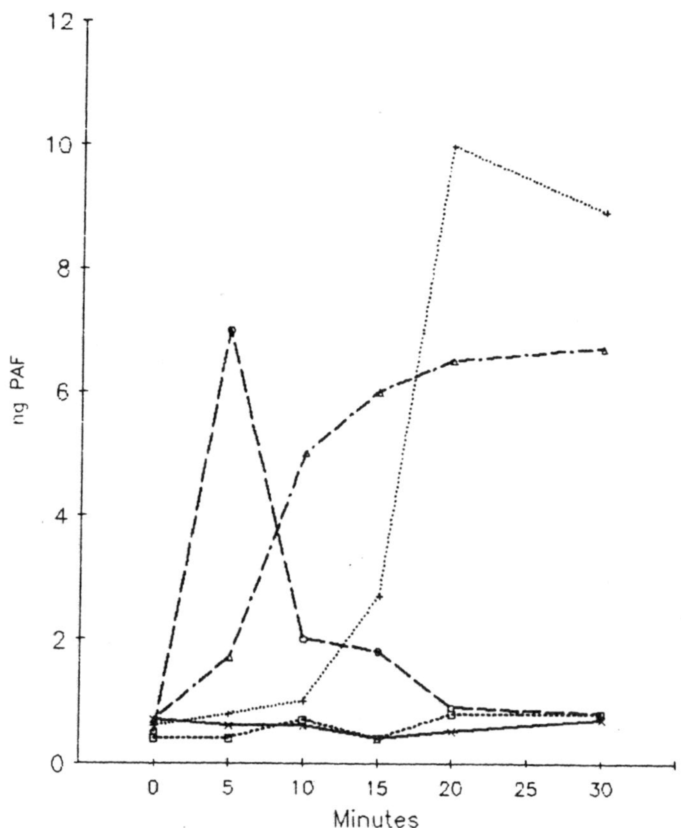

Fig. 2   PAF synthesis by unstimulated EC (□) or stimulated
         with thrombin (0.5 U/ml) (○) , angiotensin II (100 nM)
         (△) ,A23187 (1 nM) (+), FMLP (100 nM) (x). Data shown
         are representative of one of 5 experiments performed
         with similar results.

To evaluate the role of PAF produced by PMN or EC on PMN
adhesion, two PAF antagonists (BN52021 or CV3988) (6) were
tested in co-stimulatory conditions or were added to EC before
the addition of FMLP- or A23187- stimulated PMN or to PMN before the adhesion to thrombin-, angiotensin II- or A23187-
treated EC. As shown in table 3, there was a concentration-
related inhibition of PMN adhesion which, however, never
exceeded 80 % even at highest concentrations used both in co-
stimulatory and preincubation conditions. Pretreatment of EC
or PMN with 100 nM PAF, to induce selective desensitization to
this agent (12), reduced adherence by 40-70 % compared to
control (data not shown).

Table 3. Effect of BN52021 and CV3988 and PMN or EC
desensitization on PMN adhesion to EC

Condition	% inhibition (range of 3 experiments)			
	Thrombin (0.5 U/ml)	Angio II (100 nM)	FMLP (100 nM)	A23187 (1 μM)
CO-STIMULATION[a]				
BN52021 10 nM	0-3	0	0-5	0
100 nM	21-35	18-40	16-29	7-31
5 μM	61-82	34-78	20-55	19-45
10 μM	65-81	50-67	45-60	35-57
CV3988 100 nM	0	0	0	0
1 μM	18-51	23-45	12-34	20-38
10 μM	41-78	35-65	34-50	36-60
STIMULATED[b] EC TREATED WITH:				
BN52021 100 nM	N.D.	N.D.	15-27	0
5 μM	N.D.	N.D.	35-50	17-39
10 μM	N.D.	N.D.	40-57	27-49
CV3988 100 nM	N.D.	N.D.	0	0
1 μM	N.D.	N.D.	11-27	28-35
10 μM	N.D.	N.D.	20-35	26-39
STIMULATED-EC[c] PMN TREATED WITH:				
BN52021 100 nM	5-11	3-7	N.D.	N.D.
5 μM	38-61	27-60	N.D.	N.D.
10 μM	43-80	35-60	N.D.	N.D.
CV3988 100 nM	0	0	N.D.	N.D.
1 μM	11-18	17-25	N.D.	N.D.
10 μM	31-54	41-49	N.D.	N.D.

[a] In co-stimulatory experiments BN52021 or CV3988 were added 15 min before stimulation.

[b] PMN were treated with FMLP or A23187 as described in Materials and Methods and then added to EC preincubated with BN52021 or CV3988.

[c] PMN were preincubated with BN52021 or CV3988 and then added to EC stimulated with thrombin or angiotensin II as described in Materials and Methods.

Further suggestions for a role of PAF in PMN adhesion to
EC were obtained in experiments of adhesion done in the
presence of synthetic PAF. In co-stimulatory conditions, 10 nM
PAF marked increase the process of adhesion. To discriminate
whether PAF acts on PMN, or EC, or both. PAF has been used to
stimulate independently PMN or EC before the adhesion assay.
As shown in Figure 4, PAF enhanced PMN adherence to EC by
affecting the two cells.

The maximal effect was obtained by incubating EC and PMN
with 10 nM and 50 nM PAF, respectively. 1 nM PAF was active

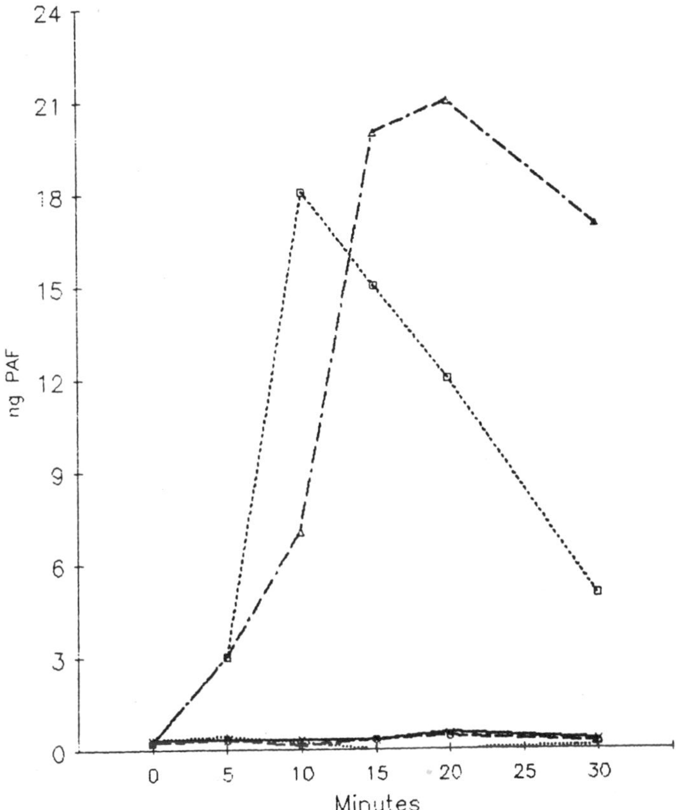

Fig. 3   PAF production by unstimulated PMN (□) , or PMN
         stimulated with thrombin (x), angiotensin II (○) ,
         FMLP (□) , or A23187 (Δ) used at the same
         concentrations of the Fig.2. Figure is representative
         of 5 experiments done with similar results.

only on EC, whereas the minimal PAF concentration activating
PMN was 5 nM (not shown). The action of PAF was rapid reaching
its maximum after 2-5 min in co-stimulatory experiments or
when added to PMN alone. In contrast, the effect of PAF on EC
was delayed reaching the maximum after 15-20 min. Stimulated
adherence persisted up to 30 min (data not shown). Lyso-PAF
(up to 1 μM) was ineffective in all conditions.

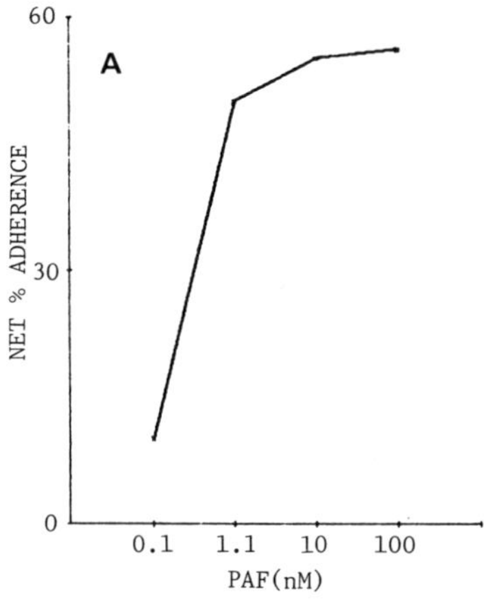

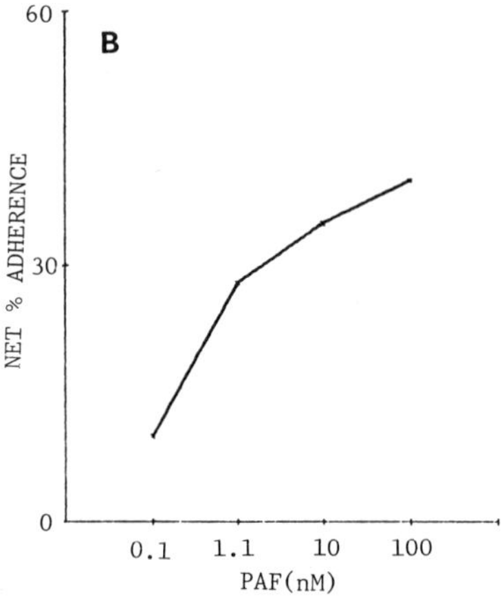

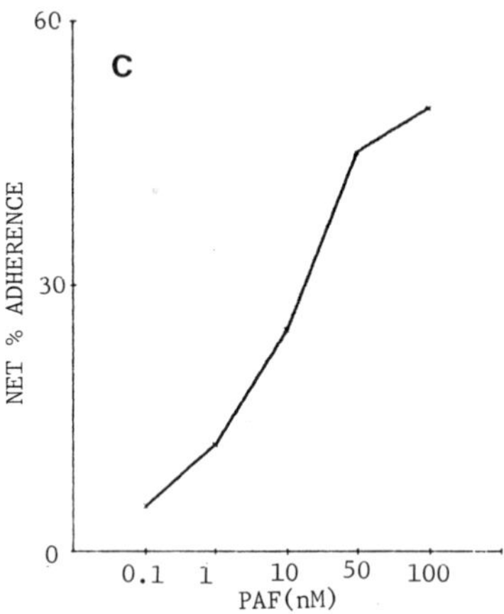

Fig. 4   Dose-dependent effect of PAF on PMN adhesion to EC.
A) Co-stimulatory condition; B) PAF-treated EC;
C:PAF-treated PMN. Comparable results were obtained
from six additional experiments.

DISCUSSION

     In this report we present evidence that PAF gives a
significant contribution in both the leukocyte-dependent and
EC-dependent mechanisms of adhesion of circulating PMN to
vasculature.

     This conclusion is argued by the following observations:
1) stimuli specific for EC (thrombin and angiotensin II)
promote the PAF synthesis only in EC and increase the adhesion
process in co-stimulatory conditions and when added to EC
before the challenge with PMN; viceversa FMLP is only active
on PMN in term of PAF production and greatly enhances the
adhesion in co-stimulatory experiments and when stimulates
PMN in the absence of EC; 2) the dependence of agonist
concentration and of the time of incubation for the synthesis
of PAF by PMN or EC strictly correlates with the adhesion
process for all stimuli examined; 3) PAF receptor antagonists
and the desensitization of both cell types to PAF partially
inhibits PMN-EC adhesion; 4) the pretreatment of PMN or EC
with PAF, but not with the biologically inactiv lyso-PAF, or
the addition of the mediator in co-stimulatory experiments
enhanced the adhesion. This activity of PAF is inhibited by
PAF receptor antagonists

     PAF is involved in an autocrine pathway of activation in
PMN as well as in EC (14). Recent reports suggest that PAF
produced by EC after thrombin stimulation represents an
alternative mechanism for PMN adherence perhaps by a molecular
alteration of EC surface (10). Furthermore, Tonnesen and
coworkers have recently demonstrated that PAF up-regulated
the expression of CD18 complex on PMN membrane (7). Our work
is consistent with these results, and suggests that PAF
produced at site of inflammed tissues have a concomitant
activity on PMN and EC representing a signal for a reciprocal
interaction.

Acknowledgement. This work was supported by AIRC.

REFERENCES

1.   Harlam, J.M.: Leukocyte-endothelial interactions. _Blood_
     65:513 (1985).
2.   Kishimoto, T.K., Jutila, M.A., Berg, E.L., Butcher, E.C.:
     Neutrophil Mac-1 and MEL-14 adhesion proteins inversely
     regulated by chemotatic factors. _Science_ 245:1238 (1989).
3.   Smith, C.W., Rothlein, R., Hughes, B.J., Mariscalco,
     M.M., Rudloff, H.E., Schmalstieg, F.C., Anderson, D.C.:
     Recognition of an endothelial determinant for CD18-
     dependent human neutrophil adherence and transendothelial
     migration. _J. Clin. Invest._ 82:1746 (1988).
4.   Luscinskas, F.W., Brock, F.A., Arnaut, M.A., Gimbrone,
     M.A.: Endothelial-leukocyte adhesion molecule-1-dependent
     and leukocyte (CD11/CD18)-dependent mechanisms contribute
     to polymorphonuclear leukocyte adhesion to cytokine-
     activated human vascular endothelium. _J. Immunol._
     142:2257 (1989).
5.   Bevilacqua, M.P., Pober, J.S., Mendrick, D.L., Cotran,

R.S., Gimbrone, M.A.: Identification of an inducible endothelial-leukocyte-adhesion molecule. <u>Proc. Natl. Acad. Sci. USA</u> 84:9238 (1987).

6. Braquet, P., Touqui, L., Shen, T.Y., Vargaftig, B.B.: Perspective in platelet activating factor research. <u>Pharmacol. Rev.</u> 39:97 (1987).

7. Tonnesen, M.G., Anderson, C.A., Spriner, T.A., Knedler, A., Avdi, N., Henson. P.M.: Adherence of neutrophils to cultured human microvascular endothelial cells. Stimulation by chemotactic peptides and lipid mediators and dependence upon the Mac-1, LFA-1 and p150,95 glyoprotein family. <u>J. Clin. Invest.</u> 83:637 (1989).

8. Kimani, G., Tonnesen, M.G., Henson, P.M.: Stimulation of eosinophils adherence to human vascular endothelial cells in vitro by platelet activating factor. <u>J. Immunol.</u> 140:3161 (1989).

9. Valone, F.H., Goetzl, E.J.: Enhancement of human polymorphonuclear leukocyte adherence by the phospholipid mediator 1-0-hexadecyl-2-acetyl-sn-glycero-3-phosphocholine (AGEPC). <u>Am. J. Pathol.</u> 113:85 (1983).

10. Zimmerman, A.G., McIntyre, T.M., Prescott, S.M.: Thrombin stimulates the adherence of neutrophils to human endothelial cells in vitro. <u>J. Clin. Invest.</u> 76:2235 (1985).

11. Camussi, G., Bussolino, F., Salvidio, G., Baglioni, C.: Tumor necrosis factor/cachectin stimulates peritoneal macrophages, polymorphonuclear neutrophils and vascular endothelial cells to synthesize and release platelet activating factor. <u>J. Exp. Med.</u> 166:1390 (1987).

12. Breviario, F., Bertocchi, F., Dejana, E., Bussolino, F.: IL-1-induced adhesion of polymorphonuclear leukocytes to cultured human endothelial cells. Role of platelet-activating factor. <u>J. Immunol.</u> 141:3391 (1988).

13. Bussolino, F., Gremo, F., Pescarmona, G.P., Camussi, G.: Platelet activating factor generation from chick retina. <u>J. Biol. Chem.</u> 261:16502 (1986).

14. Braquet, P., Paubert-Braquet, M., Bourgain, R.H., Bussolino, F., Hosford, D.: PAF/cytokine autogenerated feedback networks in microvascular immune injury: consequences in shock, ischemia and graft rejection. <u>J. Lipid Mediat.</u> 1:75 (1989).

MECHANISMS OF LIPOPOLYSACCHARIDE PRIMING FOR ENHANCED RESPIRATORY BURST ACTIVITY IN HUMAN NEUTROPHILS

J. R. Forehand, J. S. Bomalski,
R. B. Johnston, Jr.

The Children's Hospital of Philadelphia, USA

INTRODUCTION

The neutrophil is important in host defense against bacterial and fungal infections (1). That the neutrophil is capable of serving in this crucial role is made possible by virtue of its unique ability to induce microbial killing, in part, by exposing ingested microorganisms to toxic oxygen metabolites, including superoxide anion (O_2^-) - products of the respiratory burst, which commences during phagocytosis (2).

In addition to the sentinel role played in host defense, the neutrophil may contribute to tissue injury associated with the inflammatory response in conditions such as rheumatoid arthritis or immune complex glomerulonephritis (3). Therefore, the neutrophil can be viewed as a double-edged sword: beneficial in host defense against microbial invasion yet harmful if drawn into tissues during states of acute or chronic inflammation.

The reduction of oxygen to O_2^- by neutrophils is achieved by the enzymatic activity of a normally quiescent plasma membrane-associated NADPH oxidase (2). NADPH oxidase activity is triggered by a variety of soluble stimuli such as bacterial peptides or phorbol esters or by contact with opsonized particles or antigen-antibody complexes (4). Its catalytic activity can be monitored by measuring the release of O_2^- (Fig. 1) (5). The onset of O_2^- release from stimulated neutrophils is associated with a delay of 15-60 seconds (lag time), which varies with the stimulus (e.g., see Fig. 1) (4). During this interval the biochemical events occur that link cellular stimulation to the activation of the NADPH oxidase (signal transduction) (4,6).

New Aspects of Human Polymorphonuclear Leukocytes
Edited by W.H. Hörl and P.J. Schollmeyer, Plenum Press, New York, 1991

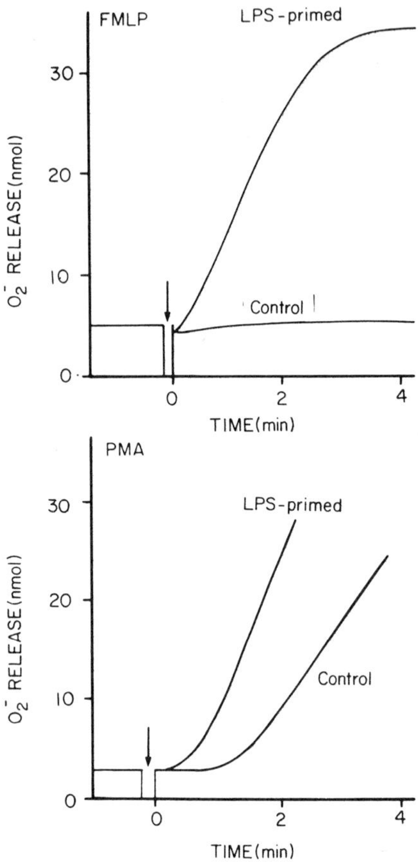

Fig. 1 The time course of O_2^- release from control and LPS-primed human neutrophils stimulated with FMLP (1 μM) (above), and PMA (100 ng/ml) (below). The tracings are drawn from a representative experiment of 25 performed in which O_2^- release from control and LPS-primed neutrophils were compared. Arrows indicate the addition of stimulus.

LIPOPOLYSACCHARIDE PRIMING

The act of preparing the neutrophil for an enhanced response to stimulation is referred to as <u>priming</u> (5,7). Neutrophils exposed to bacteria-derived lipopolysaccharide (LPS) are primed for an increased release of O_2^- in response to a variety of agonists by as much as 20-fold compared to cells not exposed to LPS (5). The release of O_2^- is not only greater but occurs sooner and at a faster rate (5,7). The effect of LPS on stimulus-induced O_2 release is both time-

and dose-dependent (7). The expression of priming requires an incubation time with LPS of at least 15 min. Neutrophils are fully primed by 75 min. Exposing cells to concentrations of LPS as low as 1 ng/ml results in an enhancement in stimulus-induced O_2^- release (7). Optimal priming occurs at LPS concentrations of 10-100 ng/ml (7). Cells exposed to LPS at 0° fail to prime, and neither extracellular Ca^{2+} nor protein synthesis is required for neutrophils to attain the primed state in response to LPS exposure (7). Changes in the characteristics of stimulus-induced O_2^- release that result from LPS priming suggest that the basis of LPS priming may be a direct effect on the NADPH oxidase, or one or more of the steps in signal transduction, or both. Defining the mechanism of LPS priming may provide investigators with a better understanding of cellular activation and with tools to regulate these events.

EFFECT OF LPS ON NADPH OXIDASE ACTIVITY OF HUMAN NEUTROPHILS

Previous studies revealed a parallel between priming of murine macrophages during the course of infection or to in vitro exposure to LPS and a change in the K_m of the NADPH oxidase for its substrate (8-10). These observed changes in enzyme kinetics were thought to represent the basis of the enhanced respiratory burst seen in association with macrophage activation, a process that is essential for cell-mediated immunity (11). We investigated the possibility that a similar mechanism might exist for LPS priming of human neutrophils. NADPH oxidase kinetics were derived from studies of NADPH-dependent O_2^- production from SDS-activated sonicates of unstimulated control or LPS-exposed neutrophils (5, 12). The kinetic parameters K_m and V_{max} were determined by fitting the experimental data to the hyperbolic curve of the Michaelis-Menten equation (5). The K_m of the NADPH oxidase-enriched fraction from control and LPS-pretreated cells was similar (19.6 ± 3.3 and 16.8 ± 2.0 M, respectively) (5). The computed V_{max} of LPS-preincubated cells (12.2 ± 8.6 nmol O_2^-/min/mg) was decreased compared to that from controls (20.2 ± 8.2 nmol O_2^-/min/mg, P < 0.05, n = 3) (5). Therefore, the basis of LPS priming for an enhanced O_2^- release could not be explained by an increase in the avidity of the NADPH oxidase or in the affinity for its substrate, NADPH.

EFFECTS OF LPS ON MEMBRANE POTENTIAL

The production of O_2^- by neutrophils in response to stimulation is associated with shifts in the intracellular concentration of various ionic species, which is mirrored by changes in transmembrane electrical potential ($\Delta\psi$) (13). The fact that changes in $\Delta\psi$ precede the release of O_2^- (13,14) and that the expression of $\Delta\psi$ and NADPH oxidase activity appear simultaneously during granulocyte development (15) raised the possibility that changes in   might be an important antecedent event to the stimulation of the NADPH oxidase. Transmembrane electric potential can be monitored by measuring the fluorescence of charged lipophilic dyes, which partition across cellular membranes in accordance with the net electri-

cal charge across those membranes (16, 17). Incubating cells
with LPS showed no apparent changes in baseline $\Delta\psi$ (5).
Nonetheless, after stimulation with FMLP the change in $\Delta\psi$ by
cells incubated with LPS was 5-fold greater than the change
in $\Delta\psi$ by control cells and paralleled the increase in O_2^-
release noted in LPS-primed neutrophils (Table I) (5).

Table I. The effect of LPS on O_2^- release and change in
membrane potential ($\Delta\psi$) in human neutrophils in response to
stimulation

Cell type	O_2^- release	$\Delta\psi$
Control	10.2 ± 5.2	7.5 ± 1.5
LPS	57.8 ± 11.9	36.9 ± 4.2
Fold difference	5.7	5.6

Suspending neutrophils in a buffer containing 50 mM KCl
effectively reduced the responsiveness of $\Delta\psi$ by 45 ± 5%, but
did prevent the expression of LPS priming as an increase in
the release of O_2^- in response to stimulation (5). Dissociating
the stimulus-induced change in $\Delta\psi$ and O_2^- release suggested
that the two events were independent and that LPS priming
affects an earlier step in signal transduction.

EFFECT OF LPS ON INTRACELLULAR FREE CALCIUM

That intracellular free Ca^{2+} plays an important role in
stimulus-induced neutrophil activity has been previously
established (18). Furthermore, elevating the concentration of
intracellular free Ca^{2+} ($[Ca^{2+}]_i$) with small amounts of the
Ca^{2+} ionophore, ionomycin, increases the release of O_2^- in
response to stimulation (19). We explored the possibility
that priming with LPS might also be associated with a rise in
$[Ca^{2+}]_i$. Neutrophils were preexposed to LPS, 50 ng/ml, for 60
min at 37°C and then loaded with the Ca^{2+}-sensitive fluorescent dye, fura-2 (20,21). The concentration of intracellular fura-2 was identical in control and LPS-primed cells
(5). When compared to control, cells preincubated with LPS
demonstrated a consistent increment (67 ± 8%, n = 12) in
resting $[Ca^{2+}]_i$ (Table II) (5). Peak $[Ca^{2+}]_i$ levels and rate
of increase in $[Ca^{2+}]_i$ in response to stimulation with FMLP
were also significantly greater. Preloading cells with
[bis-(2-amino-5-methyl-phenoxy)-ethane-N,N,N',N'-tetraacetic
acid tetraacetoxymethyl ester (MAPTAM) (22,23), a nonfluores-
cing Ca^{2+} chelator prior to LPS exposure abolished both the
incremental changes in $[Ca^{2+}]_i$ and the enhanced release of O_2^-
in response to stimulation that distinguishes LPS-priming
(Fig. 2) (5). Replenishing intracellular Ca^{2+} with ionomycin
restored neutrophils to their primed state implying that Ca^{2+}
is a crucial intracellular messenger in LPS priming (5).

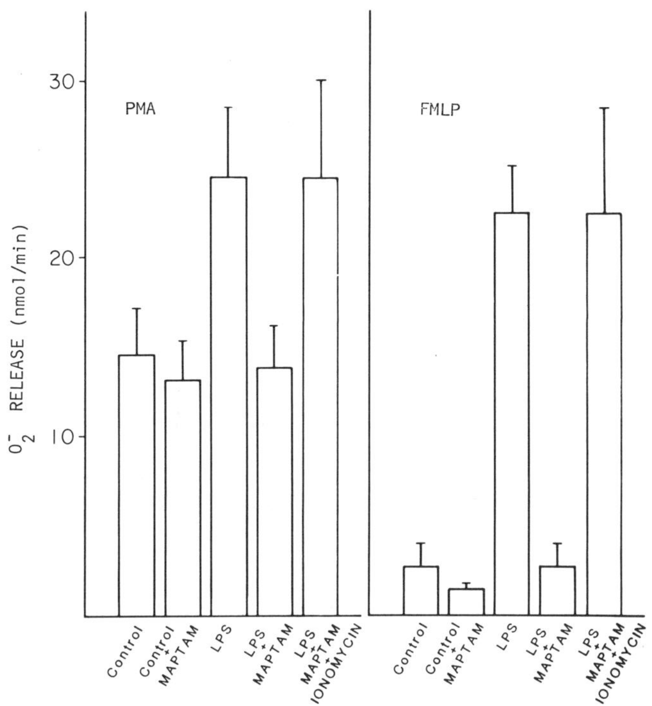

Fig. 2 The effect of the intracellular Ca^{2+} chelator MAPTAM on LPS priming. Neutrophils, preincubated in the presence or absence of MAPTAM for 30 min and then exposed to buffer or LPS, were stimulated with 100 ng/ml PMA or 1 μM FMLP. The column at the right of the panel with each stimulus represents results obtained after adding ionomycin, 1 μM, 60 s before adding the stimulus. The results are expressed as the initial rate of O_2^- release and represent the mean ± SEM of three experiments. Initial rate of O_2^- release and total O_2^- release in response to stimulation by FMLP are comparable parameters: r = 0.98, P < 0.005, n = 6 (Pearson's coeffecient of correlation). The use of rates permits more rapid analysis of cells stimulated with PMA.
(From reference 5, with permission)

Table II. The effect of LPS priming on the concentration of cytosolic free calcium ($[Ca^{2+}]_i$) and on the rate of increase in $[Ca^{2+}]_i$ in response to FMLP stimulation[*].

	Unstimulated	FMLP-stimulated			
		1nM		1μM	
		peak		peak	
Cell type	$[Ca^{2+}]_i$	$[Ca^{2+}]_i$	rate	$[Ca^{2+}]_i$	rate
	nM	nM	ΔnM/s[¶]	nM	ΔnM/s[¶]
Control	42±3	87±22	1.0±0.5	268±66	37.3±9
LPS-primed	69±5[§]	140±29[+]	5.5±1.1[+]	329±68[+]	65.0±13[+]

[*]The values reported are the result of 6 experiments for 1 uM FMLP, 5 experiments for 1 nM FMLP, and 14 experiments unstimulated.

[¶]The maximal change in $[Ca^{2+}]_i$ with time is shown.

[§]P < 0.001

[+]P < 0.05

(From reference 5, with permission)

EFFECT OF LPS ON PHOSPHOLIPASE A$_2$ ACTIVITY

Phospholipase A$_2$ enzymes are responsible for catalyzing the hydrolysis of the ester bond at the _sn_-2 position of membrane phospholipid to form a free fatty acid and lysophospholipid (24). Phospholipase A$_2$ enzymatic activity is greatly enhanced in the presence of elevated levels of intracellular Ca^{2+} (24, 25). That phospholipase A$_2$ activation from human neutrophils may play a role in priming is predicated on the findings that fatty acids and lysophospholipids, products of phospholipase A$_2$ catalytic activity, are able to enhance certain of the biochemical steps that lead to the stimulation of the NADPH oxidase (26-32). For instance, fatty acids have been shown to regulate protein kinase C activity (26), to alter membrane permeability (27, 28), and to directly trigger NADPH-dependent O$_2$ production (29). On the other hand, lysophospholipid is a biochemical intermediate in the synthesis of platelet-activating factor (PAF) (30, 31). PAF accumulates intracellularly in response to exposing neutrophils to LPS and, when added exogenously, elevates intracellular Ca^{2+} and primes neutrophils for an increase in the release of O$_2$ in response to stimulation (32).

We monitored phospholipase A$_2$ activity in neutrophils in-

cubated in buffer or buffer with LPS, 50 ng/ml. Preliminary
experiments demonstrated a consistent increase in phospholipase A_2 activity in neutrophils exposed to LPS, but not to
buffer alone. The enhanced phospholipase A_2 activity in
response to LPS was associated with the release of oleic acid,
but not arachidonic, linoleic, or palmitic acid. The LPS-me-
diated change in phospholipase A_2 activity and in priming of
neutrophils for an enhanced release of O_2^- in response to
stimulation could be equally blocked with mepacrine, a putative inhibitor of phospholipase A_2 activity (33, 34). These
results raise the possibility that phospholipase A_2 activation
is a fundamental component of priming with LPS.

CONCLUSION

To date, the results of our attempts to understand the
basis of priming with LPS point toward roles for intracellular
Ca^{2+} and phospholipase A_2 activity. The relationship of
changes in $[Ca^{2+}]_i$ and the activation of phospholipase A_2
remains to be determined. Examining the molecular basis of
LPS priming may lead to greater understanding of the mechanism
of signal transduction and those steps that might be subject
to pharmacological modification. The ability to alter signal
transduction could lead to the development of an in vivo
method to regulate (up or down) neutrophil function, and
thereby influence host defense against infection or inflammatory disorders. The results of these studies could be used to
evaluate the physiological response of a variety of secretory
cells from tissues of the nervous, endocrine or other systems,
which are also dependent on these same or similar biochemical
pathways to regulate cellular activity.

REFERENCES

1. Klebanoff, S.J., Clark, R.A.: "The Neutrophil: Function
   and Clinical Disorders", North-Holland Publishing Co.,
   New York, 1978.
2. Babior, B.M.: The respiratory burst of phagocytes. _J.
   Clin. Invest._ 73:599-601 (1984).
3. Weiss, S.J.: Tissue destruction by neutrophils. _New Engl.
   J. Med._ 320:365-376 (1989).
4. Forehand, J.R., Nauseef, W.M., Johnston, R.B. Jr.: Inherited disorders of phagocytic killing, _in_: "The Metabolic Basis of Inherited Disease", Scriver, C.R., Beaudet,
   A.L., Sly, W.S., Valle, D., eds., McGraw-Hill, New York,
   1989.
5. Forehand, J.R., Pabst, M.J., Phillips, W.A., Johnston,
   R.B. Jr.: Lipopolysaccharide priming of human neutrophils
   for an enhanced respiratory burst: Role of intracellular
   calcium. _J. Clin. Invest._ 83:74-83 (1989).
6. Korchak, H.M., Vienne, K., Rutherford, L.E., Weissmann,
   G.: Neutrophil stimulation: Receptor, membrane and
   metabolic events. _Fed. Proc._ 43:2749-2754 (1984).
7. Guthrie, L.A., McPhail, L.C., Henson, P.M., Johnston, R.B.
   Jr.: Priming of neutrophils for enhanced release of
   oxygen metabolites by bacterial lipopolysaccharides:
   Evidence for increased activity of the superoxide-producing enzyme. _J. Exp. Med._ 160:1656-1671 (1984).

8. Sasada, M., Pabst, M.J., Johnston, R.B. Jr.: Activation of mouse peritoneal macrophages by lipopolysaccharide alters the kinetic parameters of the superoxide-producing NADPH oxidase. _J. Biol. Chem._ 258:9631-9642 (1983).

9. Tsunawaki, S., Nathan, C.F.: Enzymatic basis of macrophage activation: Kinetic analysis of superoxide production in lysates of resident and activated mouse peritoneal macrophages and granulocytes. _J. Biol. Chem._ 259:4305-4312 (1984).

10. Breton, G., Cassatella, M.A., Cabrini, G., Rossi, F.: Activation of mouse macrophages causes no change in expression and function of phorbol diester receptors but is accompanied by alterations in the activity and kinetic parameters of NADPH oxidase. _Immunology_ 54:371-379 (1985).

11. Johnston, R.B. Jr.: Current concepts: Immunology - monocytes and macrophages. _New Engl. J. Med._ 318:747-752 (1988).

12. Bromberg, Y., Pick, E.: Activation of NADPH-dependent superoxide production in a cell-free system by sodium dodecyl sulfate. _J. Biol. Chem._ 260:13539-13545 (1985).

13. Korchak, H.M., Weissmann, G.: Changes in membrane potential of human granuloctyes antecedes the metabolic responses to surface stimulation. _Proc. Natl. Acad. Sci. USA_ 75:3818-3822 (1978).

14. Whitin, J.C., Chapman, C.E., Simons, E.R., Chovaniec, M.E., Cohen, H.J.: Correlation between membrane potential changes and superoxide production in human granulocytes stimulated by phorbol myristate acetate. Evidence for defective activation in chronic granulomatous disease. _J. Biol. Chem._ 255:1874-1878 (1980).

15. Kitagawa, S., Masatsugu, O., Nofiri, H., Kakinuma, K., Saito, M., Takaku, F., Mirua, Y.: Functional maturation of membrane potential changes and superoxide-producing capacity during differentiation of human granulocytes. _J. Clin. Invest._ 73:1062-1071 (1984).

16. Waggoner, A.S.: Dye indicators of membrane potential. _Ann. Rev. Biophys. Bioeng._ 8:47-68 (1979).

17. Seligmann, B.E., Gallin, J.I.: Use of lipophilic probes of membrane potential to assess human neutrophil activation. Abnormality in chronic granulomatous disease. _J. Clin. Invest._ 66:493-503 (1980).

18. Korchak, H.M., Vienne, K., Rutherfore, L.E., Wilkenfeld, C., Finkelstein, M.C., Wiessmann, G.: Stimulus response coupling in the human neutrophil. II. Temporal analysis of changes in cytosolic calcium and calcium efflux. _J. Biol. Chem._ 259:4076-4082 (1984).

19. Finkel, T.H., Pabst, M.J., Suzuki, H., Guthrie, L.A., Forehand, J.R., Phillips, W.A., Johnston, R.B. Jr.: Priming of neutrophils and macrophages for enhanced release of superoxide anion by the calcium ionophore ionomycin: Implications for regulation of the respiratory burst. _J. Biol. Chem._ 262:12589-12596 (1987).

20. Grynkiewicz, G., Poenie, M., Tsien, R.Y.: A new generation of Ca^{2+} indicators with greatly improved fluorescence properties. _J. Biol. Chem._ 260:3440-3450 (1985).

21. Korchak H.M., Vosshall, L.B., Zagon, G., Ljubich, P., Rich, A.M., Weissmann, G.: Activation of the neutrophil by calcium-mobilizing ligands. I. A chemotactic peptide and the lectin conconavalin A stimulate superoxide anion generation but elicit different calcium movement and phosphoinositide remodeling. _J. Biol. Chem._ 263:11090-11097 (1988).

22. Korchak, H.M., Vosshall, L.B., Haines, K.A., Wildenfeld, C., Lundquist, K.F., Weissmann, G.: Activation of the human neutrophil by calcium-mobilizing ligands. II. Correlation of calcium, diacyl glycerol, and phosphatidic acid generation with superoxide anion generation. <u>J. Biol. Chem.</u> 263:11098-11105 (1988).

23. Tsien, R.Y.: A non-disruptive technique for loading calcium buffers and indicators into cells. <u>Nature</u> 290:527-528 (1981).

24. Dennis, E.D.: Phospholipases, <u>in</u>: "The Enzymes", vol. XVI, Boyer, P.D. ed., Academic Press, New York, 1983.

25. Chang, J., Musser, J.H., McGregor, H.: Phospholipase A_2: Function and pharmacological regulation. <u>Biochem. Pharmacol.</u> 36:2429-2436 (1987).

26. McPhail, L.C., Clayton, C.C., Synderman, R.: A potential second messenger role for unsaturated fatty acids: Activation of Ca^{2+}-dependent protein kinase. <u>Science (Wash. DC)</u> 224:622-625 (1984).

27. Kim, D., Clapham, D.E.: Potassium channels and cardiac cells activated by arachidonic acid and phospholipids. <u>Science (Wash. DC)</u> 244:1174-1176 (1989).

28. Ordway, R.W., Walsh, J.V. Jr., Singer, J.J.: Arachidonic acid and other fatty acids directly activate potassium channels in smooth muscle cells. <u>Science (Wash.DC)</u> 244:1176-1179 (1989).

29. Curnutte, J.T., Badwey, J.A., Robinson, J.M., Karnovsky, M.J., Karnovsky, M.L.: Studies on the mechanism of superoxide release from human neutrophils stimulated with arachidonate. <u>J. Biol. Chem.</u> 259:11851-11857 (1984).

30. Stryer, L.: "Biochemistry," third edition. W. H. Freeman and Co., New York, (1988).

31. Englberger, W., Bitter-Suerman, K., Hadding, U.: Influence of lysophospholipid and PAF on the oxidative burst of PMNL. <u>Int. J. Immunolpharmacol.</u> 9:275-282 (1987).

32. Worthen, G.S., Seccombe, J.F., Clay, K.L., Guthrie, L.A., Johnston, R.B. Jr.: The priming of neutrophils by lipopolysaccharide for production of intracellular platelet-activating factor: Potential role in mediation of enhanced superoxide secretion. <u>J. Immunol.</u> 140:3553-3559 (1988).

33. Jain, M.K., Jahagirdar, D.V.: Action of phospholipase A_2 on bilayers. Effect of inhibitors. <u>Biochim. Biophys. Acta</u> 814:319-326 (1985).

34. Dennis, E.A.: Phospholipase A_2 mechanism: Inhibition and role in arachidonic acid release. <u>Drug Dev. Res.</u> 10:205 (1987).

THE EFFECT OF INFLAMMATORY MEDIATORS ON NEUTROPHIL FUNCTION

Steven D. Tennenberg

Department of Surgery
Allen Park Veterans Affairs Medical Center
Southfield and Outer Drive
Allen Park, MI 48101

INTRODUCTION

The adult respiratory distress syndrome (ARDS) is a form of acute
lung injury that most commonly afflicts patients who have suffered major
trauma or sepsis. The magnitude of this clinical problem is appreciated
in statistics that estimate there being approximately 200,000 cases per
year in the United States, costing several billion dollars in medical
expenditure. The serious nature of this problem is further amplified by
an ARDS mortality rate of 40 to 70%. Despite thirty years of intense
research we are only beginning to appreciate the pathophysiology of the
disease, prevention is nonexistent and therapeutic measures are merely
supportive (1,2).

The neutrophil (PMN) and the inflammatory process have played major
roles in our understanding of the pathogenesis of ARDS. This originally
was based on postmortem histologic studies that revealed an intense PMN
infiltration in the pulmonary parenchyma in patients who succumbed early
in the course of ARDS. It was postulated that PMN secretory products,
most notably toxic oxygen species and lysosomal enzymes, accounted for
the damage to the pulmonary alveolar-capillary membrane that character-
izes ARDS. The central role of the PMN was further strengthened by
animal studies that demonstrated an amelioration of acute lung injury in
models of ARDS when neutrophil depletion preceded the injurious insult.
The role of the inflammatory process in ARDS is based on the clinical
association of the disease to other major inflammatory illnesses most
notably sepsis, clinical studies that demonstrated manifestations of
complement activation early in the disease process, the ability to
recreate acute lung injury in animals by activating or administering
inflammatory products and the ability to ameliorate the disease with
steroid and other anti-inflammatory agent pretreatment (3-9).

In the mid 1980's, complement mediated neutrophil activation (CMNA)
was the term given to the most widely held pathogenic mechanism believed
operating in ARDS. This theory stated that known clinical conditions
that were predisposing to ARDS led to the generation of activated
complement components, most importantly C5a. The consequent
intravascular activation of PMNs and their pulmonary leukosequestration
and subsequent toxic oxygen product liberation and lysosomal
degranulation was the pivotal element that initiated the acute lung

injury. Dr. Joseph S. Solomkin, my investigative collaborator and
mentor, was an early proponent of this theory based on several clinical
and laboratory studies that he authored. Our initial clinical and ex
vivo studies, however, demonstrated that CMNA alone cannot account for
ARDS and, as suspected, other pathogenic mechanisms must be involved
(10-12).

It is the intent of this report to review observations of PMN
function and activation during acute illness and ARDS, provide in vitro
support for the ex vivo findings and finally, summarize recent
laboratory investigations that point toward the interaction of
macrophage secretory products, most notably tumor necrosis factor (TNF),
and neutrophils and speculate on the role of this interaction in host
defense, inflammation and ARDS.

METHODS

PMN isolation. Human PMNs were isolated from peripheral venous blood
of healthy adult volunteers according to the technique of Boyum with
several modifications (13). Following anticoagulation with EDTA,
leukocyte-rich plasma was obtained after Dextran T500 sedimentation.
PMNs were pelleted from the leukocyte-rich plasma by density gradient
centrifugation over Ficoll/sodium diatrizoate. Contaminating red blood
cells were removed by brief hypotonic lysis. The entire isolation was
carried out at 4°C to prevent inadvertent temperature-related PMN
activation (14,15). PMN preparations routinely contained $\geq$95% PMNs by
Wright's stain and were $\geq$98% viable by Tryptan blue exclusion.

Superoxide production. PMN superoxide (SO) anion production was
determined spectrophotometrically as the superoxide dismutase (SOD)
inhibitable reduction of ferricytochrome C (FcC). We employed a
discontinuous, total production assay and a continuous, rate-determining
assay (16,17). The discontinuous assay (10 min) used 4 x 10^{6} PMNs, 75
nmoles of FcC and 50 ug/ml of SOD. The continuous assay (5 min) used 1 x
10^{6} PMNs, 25 nmoles of FcC and 100 ug/ml of SOD. Stimulants added prior
to the 37°C incubation were n-formyl-methionyl-leucyl-phenylalanine
(FMLP, 10^{-6}M final concentration in both assays, Calbiochem, La Jolla,
CA) or phorbol myristate acetate (PMA; 10, 100 and 1000 ng/ml in the
discontinuous assay, 1 ug/ml in the continuous assay, Sigma Chemical
Co., St. Louis, MO). Spectrophotometric determinations were made with a
six-channel, microprocessor-interfaced spectrophotometer (model
Response, Gilford, Oberlin, OH). Data were calculated using an
extinction coefficient of 21,100 M^{-1} cm^{-1} (18).

FMLP receptor assay. A radioligand equilibrium binding assay with
Scatchard plot analysis was employed to determine the maximal apparent
binding capacity (B_{max}, receptors/PMN) and dissociation constant (K_{D},
nM) of FMLP receptors on whole PMNs (15,19). Sequentially added to a
series of duplicate tubes for the determination of total binding were
HBSS without calcium and magnesium, ^{3}H-FMLP (50.9 Ci/mmol, New England
Nuclear, Boston, MA) to achieve final radioligand concentrations of 1.5,
3, 6, 12, 30 and 60 nM and 2 x 10^{6} PMNs. An excess of unlabeled FMLP
(10^{-4}M) was added to a parallel set of matched tubes for the
determination of nonspecific binding. After a 30 min incubation at 4°C,
PMN bound radioligand was recovered by vacuum filtration through glass
fiber filters (Whatman GF/C) utilizing a multiple-channel cell harvester
(model M-48R, Brandel Corp., Gaithersburg, MD). Filters were washed and
counted using standard scintillation techniques.

Our initial studies attempted to delineate the status of stimulus-induced superoxide production in various disease states (20-22). When PMNs from patients with mild to moderate sepsis and no evidence of organ failure were studied, a significant increase in FMLP-induced SO production was observed (Figure 1). As anticipated, these cells also exhibited increased expression of the FMLP surface receptor (Figure 2). It is theorized that sepsis and the inflammatory products that it generates causes intravascular activation of PMNs. This activation process may be the result of PMN degranulation which results in the surface oriented mobilization of FMLP receptors. One would surmise that activated PMNs are advantageous to the host as these cells are primed for subsequent infectious challenge.

Another clinical situation that would be expected to be associated with these same manifestations is that of ARDS. Taking the theory one step further, the acute lung injury seen in ARDS is believed secondary to the liberation of SO and other toxic oxygen species. Although this process is obviously detrimental to the host, it may represent another manifestation of PMN activation. However, PMNs from patients experiencing sepsis- and trauma-related ARDS demonstrate depressed FMLP-induced oxidative capacities. As a correlate to this, PMN FMLP surface receptor expression is down-regulated (Table 1).

Table 1. Neutrophil superoxide generation and FMLP receptor status in patients with ARDS.

Patient group	FMLP SO generation	FMLP receptor status
	$(nmoles/min/10^6 \ PMN)$	receptors/PMN
Normal (n = 10)	9.6 ± 1.5	$28,600 \pm 2,500$
ARDS (n = 8)	$4.6 \pm 0.9*$	$17,700 \pm 4,200*$

*$P < 0.05$ vs normal. (Adapted from reference 20).

This finding was initially quite puzzling and stimulated our interest in finding corroborative in vitro data. We postulated that although trauma and sepsis and their resulting inflammatory stimuli could promote PMN activation, a further manifestation of this process might result in PMN down-regulation or deactivation. We further postulated that this exaggeration of the activation process might be due to further intense PMN stimulation by the same inflammatory products that initiated the activation process or that it was the result of PMN stimulation by other inflammatory products in the setting of prior activation.

IN VITRO MODEL OF GRADED NEUTROPHIL ACTIVATION

To investigate these questions, we devised an in vitro scheme that simulated graded PMN activation (22). Employing this technique, we were able to test the response of both control and septic-derived PMNs to graded, in vitro activation. Our question was whether septic-derived PMNs behaved differently than normal PMNs and whether this might explain our previous ex vivo findings.

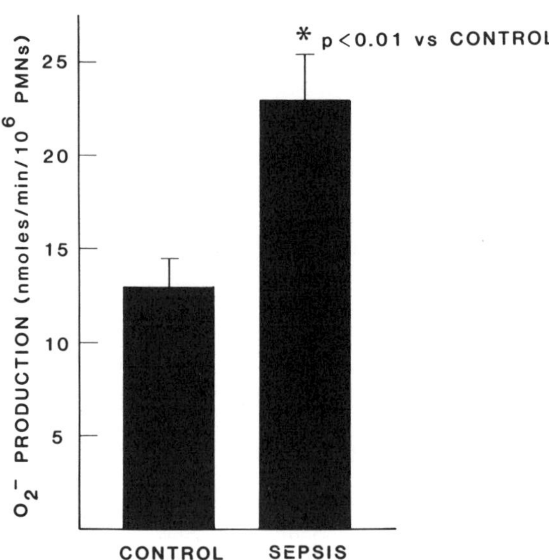

Figure 1. Neutrophil FMLP-induced SO production from controls and patients with mild to moderate sepsis. N = 8. (Adapted from reference 21).

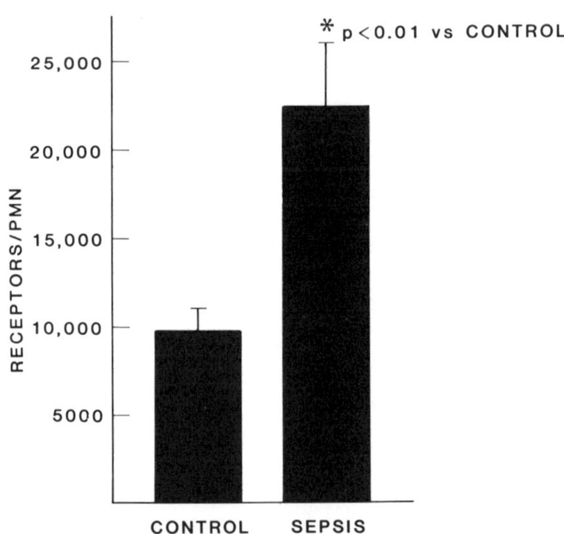

Figure 2. Neutrophil FMLP surface receptor expression from controls and patients with mild to moderate sepsis. N = 8. (Adapted from reference 21).

In the experimental protocol, PMNs from normal and septic
individuals where isolated and then subjected to either fifteen minutes
of warming at 37°C or incubation at 37°C in the presence of phorbol
myristate acetate (PMA). The control group of PMNs was maintained at
4°C. Previous studies indicated that warming simulated a moderate degree
of cellular activation, while exposure to PMA provided maximal
activation. The results of these experiments are summarized in figure 3
and table 2. Control PMNs demonstrated a graded pattern of progressive
activation with rises in FMLP-induced oxidative activity and FMLP
surface receptor expression. This was important in that it validated the
experimental model and showed that receptor expression can be used as a
measure of activation that paralleled functional activity.

Table 2. Neutrophil FMLP receptor status in control and septic-derived
PMNs following 37°C and PMA pretreatment.

	Neutrophil FMLP Receptor Status*					
	FMLP Surface Receptor Status					
	4°C Baseline		37°C Pretreatment		PMA Pretreatment	
Group	B_{max}	K_D	B_{max}	K_D	B_{max}	K_D
Control	9800 ± 1000	16.4 ± 2.7	$22\,300 \pm 3800$† (+124%)	18.4 ± 3.9	$49\,900 \pm 4000$† (+381%)	21.5 ± 2.0
Sepsis	$23\,100 \pm 2400$‡	25.5 ± 4.0	$28\,000 \pm 3400$ (+24%)‡	23.9 ± 3.1	$61\,300 \pm 7700$§ (+172%)	21.6 ± 3.9

*FMLP indicates fmet-leu-phe; PMA, phorbol myristate acetate; B_{max}, maximal apparent binding capacity; and K_D, receptor dissociation constant. Values
are the mean ± SEM. Parenthetical values represent the percent increase compared with respective 4°C baseline data.
†$P<.01$ vs 4°C baseline and each other by profile analysis.
‡$P<.001$ vs control at 4°C baseline by univariate analysis of variance and $P=.067$ vs control at 37°C pretreatment by pooled profile analysis.
§$P<.01$ vs 4°C baseline and 37°C pretreatment by profile analysis.

The response of septic-derived PMNs was, however, different. These
cells, as previously shown to be true for patients with early, mild
sepsis without organ dysfunction, were characterized by a doubling of
baseline FMLP-induced oxidative activity. In conjunction with this, FMLP
surface receptor expression was similarly increased. However, following
37°C pretreatment, whereas control cells augmented oxidative activity,
septic-derived PMNs exhibited a depression or down-regulation of FMLP-
induced oxidative activity. FMLP surface receptor expression
increased insignificantly, essentially remaining unchanged. In a pattern
and degree similar to control cells, septic PMNs demonstrated maximal
oxidative activity and receptor expression following PMA pretreatment.

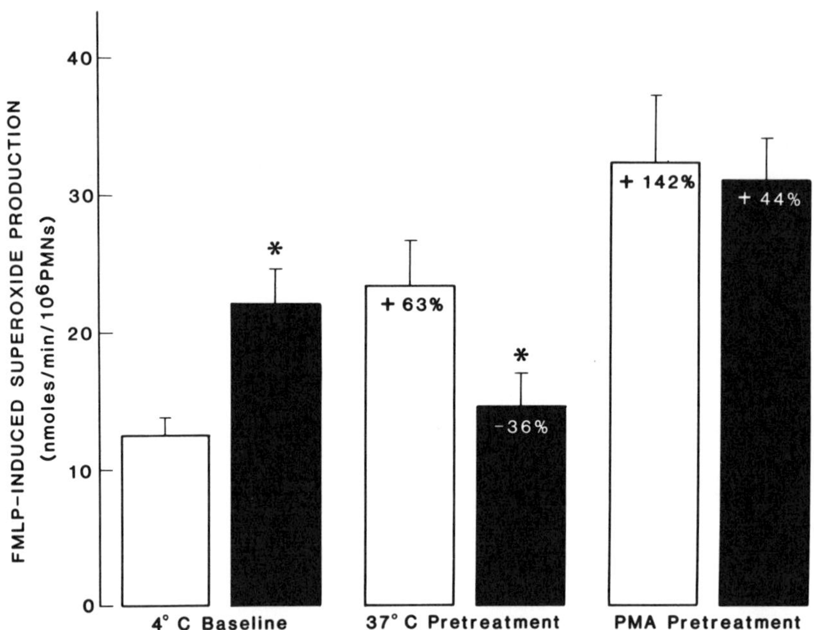

Figure 3. Effect of 37°C pretreatment and PMA pretreatment on FMLP-induced SO production by PMNs from control (open bars) and septic (closed bars) patients. N = 8. Numbers at top of bars represent percent change from the respective 4°C baseline.
*P < 0.005. (Adapted from reference 22).

The major finding of this study was in the response of septic-derived PMNs to further, moderate activation, as represented by 37°C pretreatment. These PMNs, presumably having been activated in vivo by the septic process and its mediators, demonstrated functional deactivation and down-regulation following further activation. They entered a state of apparent abnormal or dissociated stimulus-response coupling. This was evident as a maintenance of increased FMLP surface receptor expression but significantly decreased FMLP-induced oxidative activity. As demonstrated by the response to PMA incubation, the activation process, manifest by maximal receptor mobilization and oxidative priming, is intact. These observations may provide a basis for the phenomenon of PMNs from patients with septic-induced ARDS, a condition possibly indicating further PMN activation, exhibiting depressed FMLP-induced oxidative activity. In addition, these findings may provide supportive evidence for the increased septic complications, especially in the lung, following the onset of ARDS. Specifically, PMNs that had been activated in the blood and then migrate into the lung (during ARDS) and thereby undergo further activation, have depressed oxidative activity. This may translate to defective host-defense and oxidant-mediated bacterial killing in the lung, thus predisposing the ARDS patient to concurrent pneumonia.

STUDIES ON THE EFFECT OF TNF ON NEUTROPHIL FUNCTION

In order to further investigate the role of PMN activation in sepsis and possibly ARDS, we undertook a series of in vitro experiments examining the effect of cachectin/tumor necrosis factor (TNF) on PMN function (23). TNF is a macrophage-derived regulatory cytokine initially isolated in the course of studies characterizing the metabolic and tumorolytic properties of a serum factor induced in vivo by lipopolysaccharide (24-26). More recently, TNF has been implicated as an endogenous inflammatory mediator of endotoxemia and other inflammatory states, acting, in part, via stimulation of vascular endothelium and PMNs (27). The effects of TNF on endothelial cells include general activation, promotion of PMN adherence, induction of procoagulant activity and direct cytotoxicity (28-32). TNF-stimulated PMN functions include adherence, aggregation, chemotaxis, hydrogen peroxide and superoxide anion generation, degranulation, phagocytosis, fungicidal activity and antibody-dependent cellular cytotoxicity (33-40).

Macrophages and their secretory products are believed to play important roles in the regulation of local host-defense and inflammation (41-43). Amplification of these normal regulatory phenomena may also contribute to the pathogenesis of inflammation-related tissue injury as seen in septic shock, the adult respiratory distress syndrome and multiple system organ failure. Since PMN oxidative activity is necessary for effective host-defense and may contribute to tissue injury in these disease processes, we examined the role and possible mechanism(s) of the in vitro priming of PMN superoxide anion generation following exposure to TNF.

The experimental protocol for these experiments was as follows. After routine isolation, PMNs were incubated for 30 minutes at 37°C in the absence or presence of varying concentrations of TNF (purified recombinant product with minimal endotoxin contamination). After pretreatment, PMNs were pelleted by centrifugation and an aliquot of the supernatant assayed for granular enzyme release. PMNs were washed twice, recounted and assayed for SO production and FMLP receptor status.

Pretreatment of PMNs with increasing concentrations of TNF was associated with a progressive increase in FMLP-induced total SO production (Figure 4). This effect was evident at TNF concentrations as low as 10 units/ml and caused a mean 2.4-fold increase in SO production at a pretreatment concentration of 10,000 units/ml (for an E. coli-derived TNF preparation). This priming effect was most evident when a rate-determining kinetic assay of SO production was used and a 3.2-fold increase in the initial, maximal rate of SO production was observed (Figure 5).

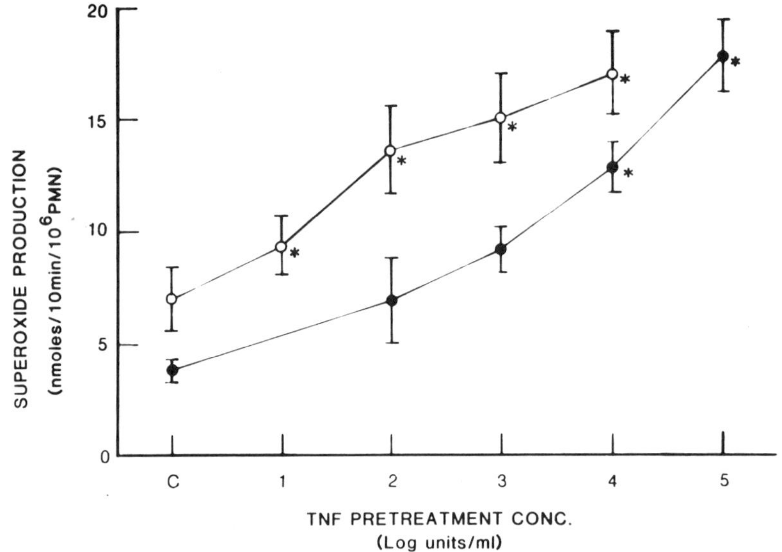

Figure 4. Effect of TNF pretreatment on FMLP-induced PMN total SO production. The effect of increasing TNF pretreatment concentrations was a statistically significant variable on SO production for both a yeast-derived (closed circles, n = 3) and E. coli-derived (open circles, n = 5) preparation. (Adapted from reference 23).

In order to determine whether this priming effect was specific to FMLP or reflective of overall cellular activation, parallel experiments were performed using PMA as the stimulant of SO production. Using a wide range of PMA concentrations and both the total and rate-determining SO assays, a priming effect was not seen following TNF pretreatment (Figures 5 and 6).

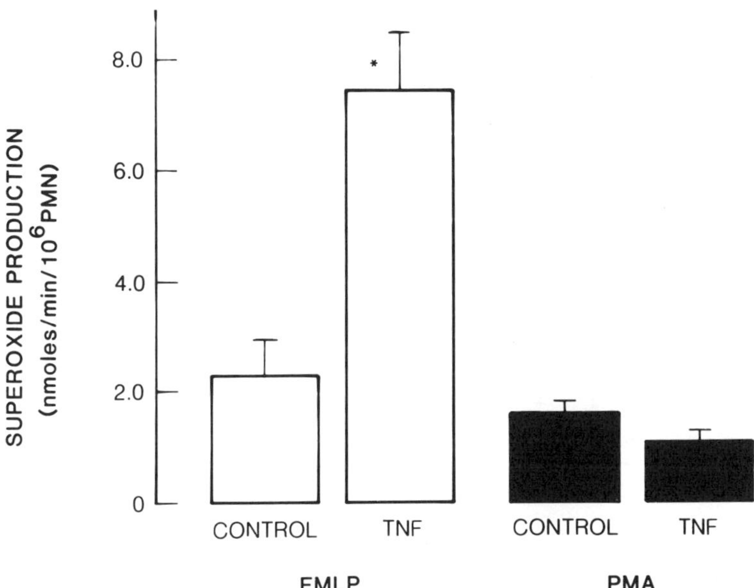

Figure 5. Effect of TNF pretreatment on FMLP- and PMA-induced SO
production. PMNs were pretreated with 10,000 units/ml TNF and the
initial, maximal rate of SO production was measured using a kinetic
assay after FMLP and PMA stimulation. N = 3 - 5. (Adapted from reference
23).

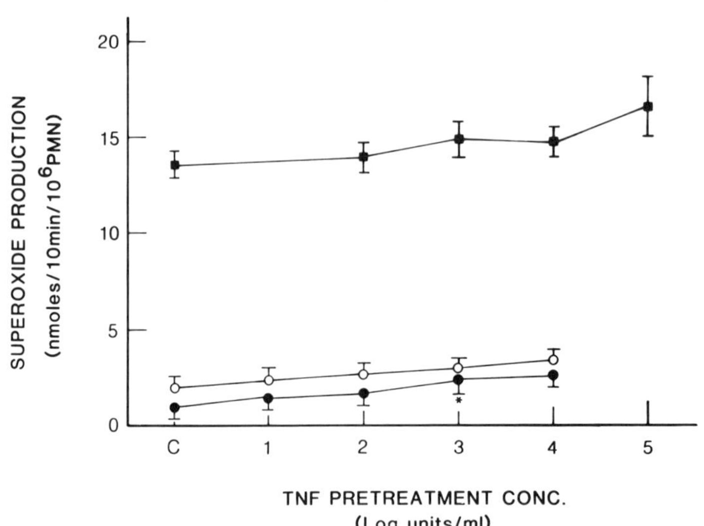

Figure 6. Effect of TNF pretreatment on PMA-induced total SO production. No priming effect was seen using PMA concentrations of 10 (closed circles, n = 2), 100 (open circles, n = 3) and 1000 (closed boxes, n = 3) ng/ml PMA. (Adapted from reference 23).

One basis for the priming effect of TNF on FMLP-induced SO production appeared to be the mobilization of PMN FMLP surface receptors. PMNs exposed to 10,000 units/ml of TNF exhibited a near doubling of FMLP surface receptor expression when compared to control cells (Table 3). Dose-response experiments showed a progressive increase in FMLP receptor expression with increasing TNF pretreatment concentrations (data not shown). The affinity of the expressed FMLP receptor was not changed, nor was the single-affinity pattern of expression (Table 3).

Table 3. FMLP surface receptor expression following TNF pretreatment.

	Yeast-derived TNF		E. coli-derived TNF	
	Control	100,000 U/ml	Control	10,000 U/ml
B_{max} (receptors/PMN)	6900 ± 2000	$11,600\pm2100*$	9800 ± 2200	$15,900\pm2500*$
K_D (nM)	21 ± 3.9	21.8 ± 4.6	16.5 ± 1.8	20.2 ± 2.5

Results are the mean of four duplicate experiments for each TNF preparation. *P < 0.01 vs. control.

Finally, we measured granular enzyme release following TNF exposure and found no increases in the release of lysozyme, B-glucuronidase or the specific granule constituent vitamin B_{12} binding protein (Figure 7).

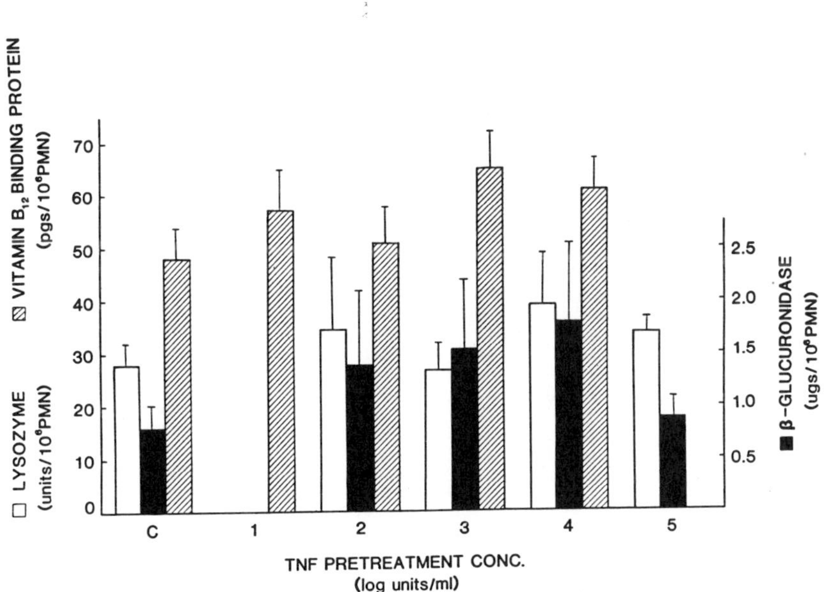

Figure 7. Effect of TNF exposure on PMN granular enzyme release. Increased release of lysozyme (open bars), B-glucuronidase (solid bars) and vitamin B_{12} binding protein (hatched bars) was not seen. The results represent the mean $\pm$ SEM of 5 - 7 experiments each assayed in duplicate. C, control. (Adapted from reference 23).

DISCUSSION

The ability of TNF to activate PMNs in vitro has fostered the hypothesis that this macrophage-derived product may play a role in in vivo bactericidal host-defense and possibly as a mediator of both the beneficial and deleterious effects of inflammation (27). Since several macrophage-derived PMN stimulants are capable of enhancing responsiveness to subsequent activation (44,45), we have explored the priming effect of TNF on PMN oxidative activity toward the chemotactic secretagogue FMLP. The potential in vivo significance of our findings lie in the fact that FMLP is released from bacteria and degenerating eukaryotic mitochondria at sites of tissue injury (46,47). Therefore, FMLP-induced PMN responses serve as a model for PMN responses toward infectious and inflammatory challenge.

Our present in vitro studies confirm that TNF, in a dose-dependent fashion, primes the subsequent oxidative responsiveness of PMNs toward FMLP. In contrast, priming was not observed with the stimulant PMA. The priming phenomenon occurred in the absence of significant degranulation and was associated with enhanced expression of FMLP surface receptors.

Several aspects of this study suggest that the mechanism for the priming phenomenon is, at least in part, an enhanced expression of PMN FMLP surface receptors. First, priming was associated with an increased expression of FMLP surface receptors without changes in receptor affinity. Since FMLP-induced responses are initiated at the PMN's surface membrane via binding to specific receptors (48,49), the increased density of FMLP receptors is likely to contribute toward a primed response. Our data also suggest that factors other than receptor number alone contribute to the oxidative response (50). This can be inferred from the lack of a one to one relationship between absolute FMLP receptor expression and oxidative activity. For the E. coli-derived TNF preparation, receptor expression increased 1.6-fold while the initial, maximal rate of primed superoxide production increased 3.2-fold. The absence of a change in receptor affinity may serve to preclude a nonspecific membrane effect as the cause of the priming phenomenon.

Berkow et al. and Atkinson et al. have both recently demonstrated the priming effect of TNF on PMN FMLP-induced oxidative responsiveness (51,52). Berkow et al., in a limited fashion, showed that this priming effect was associated with a 30% increase in PMN FMLP receptor expression. Being statistically insignificant, this was not felt to account for the more pronounced priming effect. Atkinson and coworkers found no change in absolute receptor number but did find a shift from a dual to a single-affinity model of FMLP receptor expression following TNF exposure. They proposed that TNF-induced changes in receptor affinity may play a role in the priming effect. Subtle differences in experimental technique (PMN isolation technique, incubation time, incubation media, receptor assay technique) may account for the lack of uniformity between these findings and our own.

Also supporting a role for FMLP receptor mobilization in the TNF priming response is the finding that priming was not observed with the stimulant PMA. This observation is in agreement with the recent work of Yuo et al. (53). This eliminates an effect on the post-receptor transductional components of FMLP-induced oxidative activity, at least

to the extent assessed by PMA, as a mechanism for TNF-mediated oxidative
priming. These transductional components include activity or expression
of intramembranous and cytosolic PMA receptors and the role of
diacylglycerols in the stimulation of protein kinase C (54,55).

The dissociation of oxidative priming toward various stimuli is not
unique to TNF. Another cytokine, granulocyte-macrophage colony
stimulating factor (GM-CSF), has recently been shown to prime PMN
oxidative responses toward FMLP and not PMA (56,57). Sullivan et al. has
demonstrated that the priming effect of GM-CSF on FMLP-induced SO
production is unrelated to translocation of protein kinase C or
degranulation, but was associated with release of arachidonic acid from
plasma membrane phospholipids (58). Therefore, GM-CSF and TNF may prime
PMNs in a similar fashion since GM-CSF has also been shown to increase
expression of PMN FMLP surface receptors (56,59). Platelet activating
factor is another macrophage-derived inflammatory mediator that we have
also observed to prime PMN oxidative responses toward FMLP and not PMA,
as well as mobilize FMLP surface receptors (unpublished observations).

Gallin et al. has proposed that increased FMLP receptor
expression underlies the phenomenon of FMLP-associated response priming
(60,61). It is theorized that surface-oriented mobilization of an
intracellular pool of FMLP receptors occurs secondary to degranulation
of specific granules and subsequent fusion of receptor-rich granular
membrane with the surface plasma membrane. Although TNF has been shown
to cause PMN release of granular enzymes under mild lytic conditions
(35), we were unable to demonstrate an increase in granular enzyme
release, compared to controls, with our pretreatment protocol. Most
specifically, TNF pretreatment did not result in specific granule
release as evident from the vitamin B_{12} binding protein (and lysozyme)
release data (Figure 7). Since the PMN's specific granules are thought
to contain an intracellular pool of translocatable FMLP receptors, their
lack of involvement in the FMLP receptor up-regulation phenomenon that
accompanies TNF pretreatment would suggest that another pool of receptor
material is involved. Our data show that TNF, presumably following
binding to its surface receptor (36,62), is capable of increasing PMN
FMLP receptor expression via a mechanism not involving degranulation.
This may involve uncovering of cryptic intramembranous receptors
secondary to a direct membrane effect or mobilization of as yet
undefined intracellular receptor pools.

Tissue macrophages, via release of TNF, may therefore
potentiate PMN oxidative host-defense and inflammatory responses by
priming PMNs that have been recruited to sites of infection or
inflammation (Figure 8). The recent observation of increased FMLP
receptor expression without affinity changes on rabbit PMNs following
endotoxin injection may provide in vivo corroboration of our findings
(63). This is evident since TNF has been shown to be an important
mediator of the effects of endotoxin in rabbits (27,64). The lung
represents one organ that is a likely site for TNF-mediated PMN
oxidative priming to occur. Alveolar macrophages, via production of TNF
into the lungs' interstitium (65), may prime the oxidative
responsiveness of recruited PMNs. Although this process may be involved
in normal PMN oxidative functions, it may also provide a basis for
oxidative tissue injury as seen in septic shock, adult respiratory
distress syndrome and multiple system organ failure (66,67).

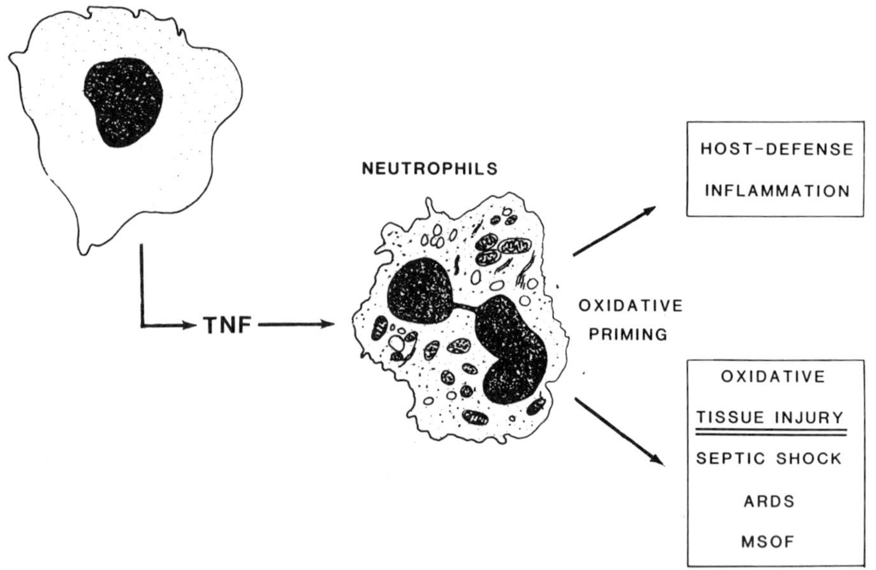

Figure 8. Proposed role of macrophage-derived TNF in the inflammatory process.

REFERENCES

1. S. H. Norwood, J. M. Civetta. The adult respiratory distress syndrome. Surg Gynecol Obstet 161:497, 1985.
2. N. Andreadis, T. L. Petty. Adult respiratory distress syndrome: Problems and progress. Am Rev Respir Dis 132:1344, 1985.
3. J. E. Rinaldo, R. M. Rogers. Adult respiratory distress syndrome: Changing concepts of lung injury and repair. N Engl J Med 306:900, 1982.
4. K. L. Brigham. Mechanisms of lung injury. Clin Chest Med 3:9, 1982.
5. R. M. Tate, J. E. Repine. Neutrophils and the adult respiratory distress syndrome. Am Rev Respir Dis 128:552, 1983.
6. P. R. Craddock. Complement, granulocytes and shock lung. Am J Emerg Med 2:78, 1983.
7. D. E. Hammerschmidt, L. J. Weaver, L. D. Hudson, et al. Association of complement activation and elevated plasma-C5a with adult respiratory distress syndrome. Lancet 1:947, 1980.
8. A. C. Heflin, K. L. Brigham. Granulocyte depletion attenuates increased lung vascular permeability after endotoxemia in sheep. J Clin Invest 68:1253, 1981.
9. K. L. Brigham, R. E. Bowers, C. R. McKeen. Methylprednisolone prevention of increased lung vascular permeability following endotoxemia in sheep. J Clin Invest 67:1103, 1981.
10. J. S. Solomkin, L. A. Cotta, P. S. Satoh, et al. Complement activation and clearance in acute illness and injury: Evidence for C5a as a cell-directed mediator of the adult respiratory distress syndrome in man. Surgery 97:668, 1985.
11. J. S. Solomkin, L. A. Cotta, J. D. Ogle, et al. Complement-induced expression of cryptic receptors on the neutrophil surface: A mechanism for regulation of acute inflammation in trauma. Surgery 96:336, 1984.

12. S. D. Tennenberg, M. P. Jacobs, J. S. Solomkin. Complement-mediated
    neutrophil activation in sepsis- and trauma-related adult
    respiratory distress syndrome: Clarification with radioaerosol
    lung scans. Arch Surg 122:26, 1987.
13. A. Boyum. Isolation of mononuclear cells and granulocytes from human
    blood. Scand J Clin Lab Invest 21 (suppl 97):77, 1968.
14. D. T. Fearon, L. A. Collins. Increased expression of C3b receptors
    on polymorphonuclear leukocytes induced by chemotactic factors
    and by purification procedures. J Immunol 130:370, 1983.
15. S. D. Tennenberg, F. P. Zemlan, J. S. Solomkin. Characterization of
    n-formyl-methionyl-leucyl-phenylalanine receptors on human
    neutrophils: Effects of isolation and temperature on receptor
    expression and functional activity. J Immunol 141:3937, 1988.
16. I. M. Goldstein, D. Roos, H. B. Kaplan, et al. Complement and
    immunoglobulins stimulate superoxide production by human
    leukocytes independently of phagocytosis. J Clin Invest
    56:1155,1975.
17. J. J. Zimmerman, J. H. Shelhamer, J. E. Parrillo. Examination of the
    enzyme assay for NADPH oxidoreductase; application to
    polymorphonuclear leukocyte superoxide anion generation. Crit
    Care Med 13:197, 1985.
18. V. Massey. The microestimation of succinate and the extinction
    coefficient of cytochrome c. Biochim Biophys Acta 34:255, 1959.
19. L. T. Williams, R. Snyderman, M. C. Pike, et al. Specific receptor
    sites for chemotactic peptides on human polymorphonuclear
    leukocytes. Proc Natl Acad Sci USA 74:1204, 1977.
20. S. D. Tennenberg, F. P. Zemlan, J. S. Solomkin. Altered neutrophil
    oxidative metabolism and receptor status in sepsis- and trauma-
    related adult respiratory distress syndrome. Surg Forum
    37:98,1986.
21. S. D. Tennenberg, J. S. Solomkin. Increased expression of neutrophil
    chemoattractant receptors in sepsis: Correlation with functional
    activation. Curr Surg 44:313, 1987.
22. S. D. Tennenberg, J. S. Solomkin. Neutrophil activation in sepsis:
    The relationship between fmet-leu-phe receptor mobilization and
    oxidative activity. Arch Surg 123:171, 1988.
23. S. D. Tennenberg, J. S. Solomkin. Activation of neutrophils by
    cachectin/tumor necrosis factor: Priming of N-formyl-methionyl-
    leucyl-phenylalanine-induced oxidative responsiveness via
    receptor mobilization without degranulation. J Leuk Biol (in
    press).
24. E. A. Carswell, L. J. Old, R. L. Kassel, et al. An endotoxin-induced
    serum factor that causes necrosis of tumors. Proc Natl Acad Sci
    USA 72:3666, 1975.
25. M. Kawakami, P. H. Pekala, M. D. Lane, et al. Lipoprotein lipase
    suppression in 3T3-L1 cells by an endotoxin-induced mediator
    from exudate cells. Proc Natl Acad Sci USA 79:912, 1982.
26. B. Beutler, J. Mahoney, N. Le Trang, et al. Purification of
    cachectin, a lipoprotein lipase-suppressing hormone secreted by
    endotoxin-induced RAW 264.7 cells. J Exp Med 161:984, 1985.
27. B. Beutler, A. Cerami. Cachectin: More than a tumor necrosis factor.
    N Engl J Med 316:379, 1987.
28. M. P. Bevilacqua, J. S. Pober, J. R. Majeau, et al. Recombinant
    tumor necrosis factor induces procoagulant activity in cultured
    human vascular endothelium: Characterization and comparison with
    the actions of interleukin 1. Proc Natl Acad Sci USA 83:4533,
    1986.
29. P. P. Nawroth, D. M. Stern. Modulation of endothelial cell
    hemostatic properties by tumor necrosis factor. J Exp Med
    163:740, 1986.

30. T. H. Pohlman, K. A. Stanness, P. G. Beatty, et al. An endothelial cell surface factor(s) induced in vitro by lipopolysaccharide, interleukin 1, and tumor necrosis factor-α increases neutrophil adherence by a CDw18-dependent mechanism. J Immunol 136:4548, 1986.

31. J. S. Pober, M. A. Gimbrone Jr, L. A. Lapierre, et al. Overlapping patterns of activation of human endothelial cells by interleukin 1, tumor necrosis factor, and immune interferon. J Immunol 137:1893, 1986.

32. N. Sato, T. Goto, K. Haranaka, et al. Actions of tumor necrosis factor on cultured vascular endothelial cells: Morphologic modulation, growth inhibition, and cytotoxicity. J N C I 76:1113, 1986.

33. J. Y. Djeu, D. K. Blanchard, D. Halkias, et al. Growth inhibition of Candida albicans by human polymorphonuclear neutrophils: Activation by interferon-γ and tumor necrosis factor. J Immunol 137:2980, 1986.

34. J. R. Gamble, J. M. Harlan, S. J. Klebanoff, et al. Stimulation of the adherence of neutrophils to umbilical vein endothelium by human recombinant tumor necrosis factor. Proc Natl Acad Sci USA 82:8667, 1985.

35. S. J. Klebanoff, M. A. Vadas, J. M. Harlan, et al. Stimulation of neutrophils by tumor necrosis factor. J Immunol 136:4220, 1986.

36. J. W. Larrick, D. Graham, K. Toy, et al. Recombinant tumor necrosis factor causes activation of human granulocytes. Blood 69:640, 1987.

37. W. J. Ming, L. Bersani, A. Mantovani. Tumor necrosis factor is chemotactic for monocytes and polymorphonuclear leukocytes. J Immunol 138:1469, 1987.

38. B. Perussia, M. Kobayashi, M. E. Rossi, et al. Immune interferon enhances functional properties of human granulocytes: Role of receptors and effect of lymphotoxin, tumor necrosis factor, and granulocyte-macrophage colony stimulating factor. J Immunol 138:765, 1987.

39. M. R. Shalaby, B. B. Aggarwal, E. Rinderknecht, et al. Activation of human polymorphonuclear neutrophil functions by interferon-γ and tumor necrosis factor. J Immunol 135:2069, 1985.

40. M. Tsujimoto, S. Yokota, J. Vilcek, et al. Tumor necrosis factor provokes superoxide anion generation from neutrophils. Biochem Biophys Res Comm 137:1094, 1986.

41. C. F. Nathan, H. W. Murray, Z. A. Cohn. The macrophage as an effector cell. N Engl J Med 303:622, 1980.

42. J. E. Pennington, T. H. Rossing, L. W. Boerth, et al. Isolation and partial characterization of a human alveolar macrophage-derived neutrophil-activating factor. J Clin Invest 75:1230, 1985.

43. R. Takemura, Z. Werb. Secretory products of macrophages and their physiologic functions. Am J Physiol 246:C1, 1984.

44. J. G. Bender, L. C. McPhail, D. E. Van Epps. Exposure of human neutrophils to chemotactic factors potentiates activation of the respiratory burst enzyme. J Immunol 130:2316, 1983.

45. J. C. Gay, J. K. Beckman, K. A. Zaboy, et al. Modulation of neutrophil oxidative responses to soluble stimuli by platelet activating factor. Blood 67:931, 1986.

46. H. Carp. Mitochondrial N-formylmethionyl proteins as chemoattractants for neutrophils. J Exp Med 155:264, 1982.

47. W. A. Marasco, S. H. Phan, H. Krutzsch, et al. Purification and identification of formyl-methionyl-leucyl-phenylalanine as the major peptide neutrophil chemotactic factor produced by Escherichia coli. J Biol Chem 259:5430, 1984.

48. R. Snyderman, M. C. Pike. Chemoattractant receptors on phagocytic cells. Ann Rev Immunol 2:257, 1984.

49. L. T. Williams, R. Snyderman, M. C. Pike, et al. Specific receptor sites for chemotactic peptides on human polymorphonuclear leukocytes. Proc Natl Acad Sci USA 74:1204, 1977.

50. L. A. Sklar, A. J. Jesaitis, R. G. Painter. The neutrophil N-formyl peptide receptor: Dynamics of ligand receptor interactions and their relationship to cellular responses. In Contemporary Topics in Immunobiology, vol 14. R. Snyderman, ed. Plenum Press, NY, p. 29, 1984.

51. Y. H. Atkinson, W. A. Marasco, A. F. Lopez, et al. Recombinant human tumor necrosis factor. Regulation of N-formylmethionylleucyl-phenylalanine receptor affinity and function on human neutrophils. J Clin Invest 81:759, 1988.

52. R. L. Berkow, D. Wang, J. W. Larrick, et al. Enhancement of neutrophil superoxide production by preincubation with recombinant human tumor necrosis factor. J Immunol 139:3783, 1987.

53. A. Yuo, S. Kitagawa, I. Suzuki, et al. Tumor necrosis factor as an activator of human granulocytes. Potentiation of the metabolisms triggered by the Ca^{2+}-mobilizing agents. J Immunol 142:1678, 1989.

54. M. Y. Castagna, Y. Takai, K. Kaibuchi, et al. Direct activation of calcium-activated, phospholipid-dependent protein kinase by tumor-promoting phorbol esters. J Biol Chem 257:7847, 1982.

55. J. Nishihira, J. T. O'Flaherty. Phorbol myristate acetate receptors in human polymorphonuclear neutrophils. J Immunol 135:3439, 1985.

56. D. English, H. H. Broxmeyer, T. G. Gabig, et al. Temporal adaptation of neutrophil oxidative responsiveness to n-formyl-methionyl-leucyl-phenylalanine. Acceleration by granulocyte-macrophage colony stimulating factor. J Immunol 141:2400, 1988.

57. R. H. Weisbart, L. Kwan, D. W. Golde, et al. Human GM-CSF primes neutrophils for enhanced oxidative metabolism in response to the major physiological chemoattractants. Blood 69:18, 1987.

58. R. Sullivan, J. D. Griffin, E. R. Simons, et al. Effects of recombinant human granulocyte and macrophage colony-stimulating factors on signal transduction pathways in human granulocytes. J Immunol 139:3422, 1987.

59. R. H. Weisbart, D. W. Golde, J. C. Gassen. Biosynthetic human GM-CSF modulates the number and affinity of neutrophil f-met-leu-phe receptors. J Immunol 137:3584, 1986.

60. M. P. Fletcher, B. E. Seligmann, J. I. Gallin. Correlation of human neutrophil secretion, chemoattractant receptor mobilization, and enhanced functional capacity. J Immunol 128:941, 1982.

61. W. Zimmerli, B. Seligmann, J. I. Gallin. Exudation primes human and guinea pig neutrophils for subsequent responsiveness to the chemotactic peptide N-formylmethionyl-leucylphenylalanine and increases complement component C3bi receptor expression. J Clin Invest 77:925, 1986.

62. M. R. Shalaby, M. A. Palladino Jr, S. E. Hirabayashi, et al. Receptor binding and activation of polymorphonuclear neutrophils by tumor necrosis factor-alpha. J Leuk Biol 41:196, 1987.

63. D. W. Goldman, H. Enkel, L. A. Gifford, et al. Lipopolysaccharide modulates receptors for leukotriene B_4, C5a, and formyl-methionyl-leucyl-phenylalanine on rabbit polymorphonuclear leukocytes. J Immunol 137:1971, 1986.

64. B. A. Beutler, I. W. Milsark, A. Cerami. Cachectin/tumor necrosis factor: Production, distribution, and metabolic fate in vivo. J Immunol 135:3972, 1985.

65. E. A. Rich, J. R. Panuska, C. B. Wolf, et al. Potent production of tumor necrosis factor-alpha by human alveolar macrophages (abstract). Clin Res 35:632A, 1987.

66. K. J. Tracey, B. Beutler, S. F. Lowry, et al. Shock and tissue
    injury induced by recombinant human cachectin. Science 234:470,
    1986.
67. K. J. Tracey, S. F. Lowry, T. J. Fahey III, et al. Cachectin/tumor
    necrosis factor induces lethal shock and stress hormone
    responses in the dog. Surg Gynecol Obstet 164:415, 1987.

CLINICAL EVALUATION OF HEMATOPOIETIC GROWTH FACTORS

A. Lindemann, F. Herrmann, R. Mertelsmann

Department I Internal Medicine,
Albrecht-Ludwigs-Universität Freiburg, FRG

INTRODUCTION

The process of formation of polymorphonuclear neutrophils
(PMN) and other peripheral blood cells (PBC) from their respective progenitors has been extensively studied over the
last few years. In in vitro culture systems using semi-solid
media it has been shown that a hierarchy of growth factors including interleukin-3 (IL-3), granulocyte-macrophage colony-

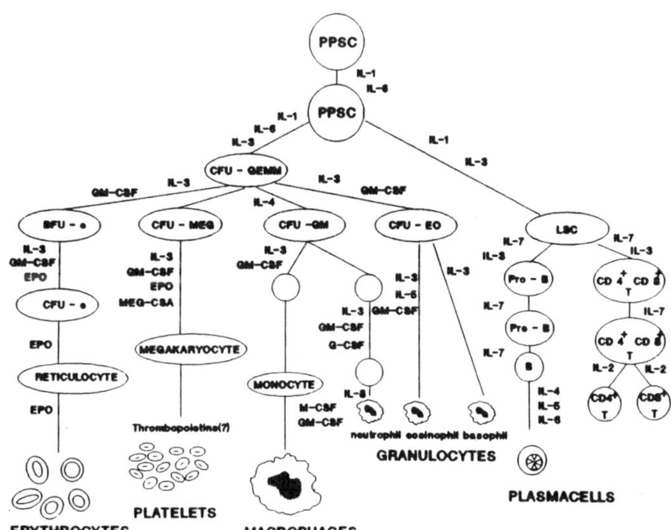

Figure 1. Differentiation of pluripotent stem cells (PPSC)
via different commited progenitor cells (colony forming unit,
CFU) of lymphoid (lymphoid stem cell, LSC), megacaryocyte
(MEG), erythrocyte (e), eosinophil (Eo), and granulocyte-
macrophage (GM) specificity into functional end-stage cells
and its modulation by various growth factors.

stimulating factor (GM-CSF), granulocyte colony-stimulating
factor (G-CSF), erythropoietin (EPO) modulate proliferation of
hematopoietic progenitors and differentiation into functionally competent and stage cells of their respective lineage
(1) (Fig. 1).

At present it is not clear, if steady state hematopoiesis
also depends on the presence of these cytokines, while they
are obviously neccessary to spur hematopoiesis in states of
increased demand, such as inflammatory injury or infections.
Up to now the clinical evaluation of the hematopoietic growth
factors (HGFs) has concentrated on EPO, IL-3, G-CSF and GM-
CSF. Our first experience with these molecules in phase I/II
studies will be reviewed below.

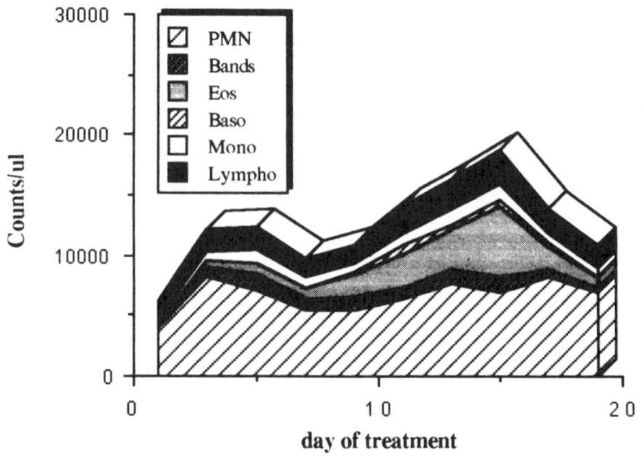

Figure 2. White blood cell counts and differential (median)
of three patients treated with IL-3 125 μg/m^2 day 1-15.

The clinical value of EPO administration is already well
established in patients with anemia of renal origin (2). The
effectivity has been regarded as the result of substitution of
a hormone deficit since these patients have low to undetec-
table endogenous erythropoietin levels. However, an increase
of the red blood cell mass can also be seen in a number of patients with other forms of anemias with normal or even elevated EPO serum levels, may be chemotherapy-induced, due to bone
marrow infiltration by neoplastic cells or dyserythropoiesis
(3,4). It seems that the feedback increase of endogenous EPO
levels in those patients responding to exogenous EPO may be
inadequate in order to exploit the full repertoire of erythro-
poiesis that may be induced, however, by supra-physiological
or pharmacological doses of the hormone. The application of
CSFs in chemotherapy induced leukopenia probably represents a
similar approach. Adverse effects of EPO administration were
not observed in this study, although the risk of cardio-
vascular complications was found to be increased in patients
with renal failure in previous studies (2). Restriction of

these side effects, however, to the latter group argues for an
indirect effect by the increased hemoglobin levels as far as
these patients are incapable to rapidly adjust to normal
hemodynamic parameters because of a possibly predamaged
vasculary system. Thus these adverse events were not attributable to the hormone itself.

The most multifunctional of the HGF, IL-3, has recently
been entered into clinical trials. Based on in vitro studies
(5), stimulation of various hematopoietic lineages was expected. The initial phase-I evaluation was conducted in a two
center study of our Department in cooperation with the Department Hematology in Frankfurt (6,7). IL-3 was administered sub-
cutaneously for 15 days to patients assigned for later chemotherapy. A minor increase of neutrophils was observed after 3
days, however, the essential increase of neutrophils and
eosinophils did not occur before 10 days of treatment
(Fig. 2). Basophils and lymphocyte counts were only slightly
elevated. The delayed onset of the hematopoietic response was
quite different from our experience with G- or GM-CSF
(Fig. 4) and suggested a stimulatory effect of IL-3 on very
early hematopoietic cells (Fig. 2.)

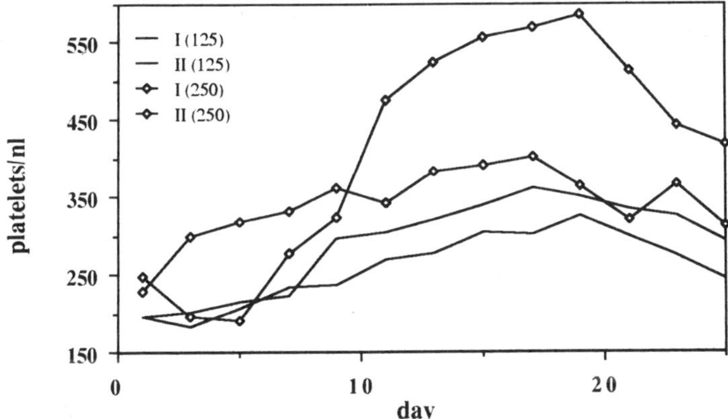

Figure 3. Platelet counts of two patients each treated at
125 μg/m^2 and 250 μg/m^2 of IL-3, day 1-15.

The most interesting effect of IL-3 was related to the
stimulation of thrombopoiesis. About two thirds of the patients treated at higher doses experienced a two- to four-fold
increase of the platelet count, that occured quite late on
therapy and was sustained in several cases for another 2-3
weeks (Fig. 3). We also observed a slight but not dose-
dependent stimulatory effect on reticulocyte counts. Although
this did not result in increased hematocrit levels, it may

qualify IL-3 as a synergizing agent with EPO for the therapy
of refractory anemia. - Adverse clinical effects were generally mild, resembling a flue like syndrome with fever, headache, weakness and sometimes local reactions at sites of
administration. This profile was quite similar to GM-CSF related side effects suggesting that IL-3 is capable of inducing
inflammatory responses like GM-CSF, may be by induction of
monocyte IL-1 or TNF-alpha secretion (8).

The granulocyte colony-stimulating factor <u>G-CSF</u> is a more
specifically acting HGF that induces the amplification of late
neutrophilic progenitors. As early as 6 hours after i.v. administration with blood cell (WBC) counts increased (9), suggesting a rapid release of mature cells from the bone marrow
reserve pool. In contrast to GM-CSF (10), once daily i.v. administration was sufficient to maintain a dose dependent elevation of WBC counts (Fig. 4). Dose escalation in the phase-I
study was stopped at 60 μg/kg because the pre-arranged limit
of 75.000 cells/μl was reached, not because of drug related
side effects.

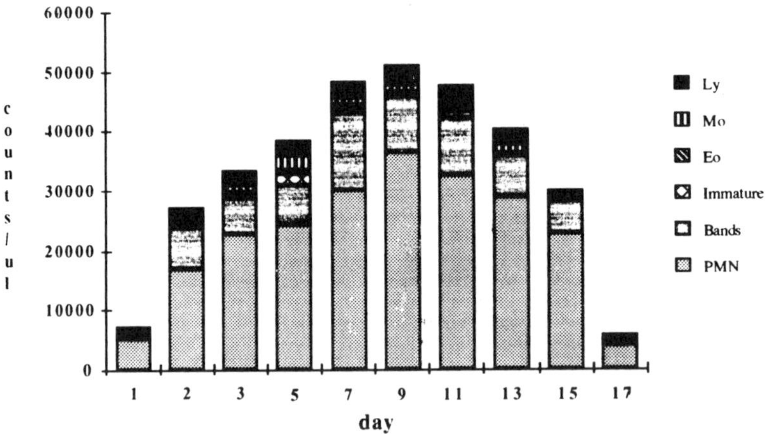

Figure 4. Total white blood cell counts and differential of
three patients (median) treated with G-CSF 30 μg/kg, day 1-14.

As evidenced from the differential count, leukocytosis
was made up by mature neutrophils and band forms almost exclusively (Fig. 4). The immediate effect of G-CSF administration was a transient drop of WBC counts, similarly observed
after GM-CSF or TNF-alpha injection. While lymphocytes and
eosinophils, lacking G-CSF receptors, were unaffected,
neutrophils and monocytes disappeared during the first 30
minutes probably due to margination to the vessel wall. We
were unable to detect a modulation of the CD11 clustered
adhesion molecules that has been reported for GM-CSF (9,10).
Yet it is unknown what really happens to the circulating
phagocytes immediately after administration of the GF.

96

Another event that remains to be elucidated concerns a
dose dependent thrombocytopenia of mild to moderate degree,
that was most prominent after about 10 days and resolved
spontaneously despite continuation of therapy (Fig. 5). An
unimpaired thrombopoiesis in the bone marrow, an increase of
platelet volume and platelet factor 4 serum levels argue for
an enhanced turnover of platelets, also supported by a
certain rebound after treatment cessation.

Laboratory changes detected in the serum paralleled elevation of WBC counts, reflecting an enhanced cellular turnover (uric acid) and a probably non-specific leakage of intracellular enzymes such as elastase and lysozyme. Serum levels
of alkaline phosphatase (bone specific isoenzyme) and lactate-
dehydrogenase (isoenzyme 4 and 5) were also highly elevated in
the responding patients (9).

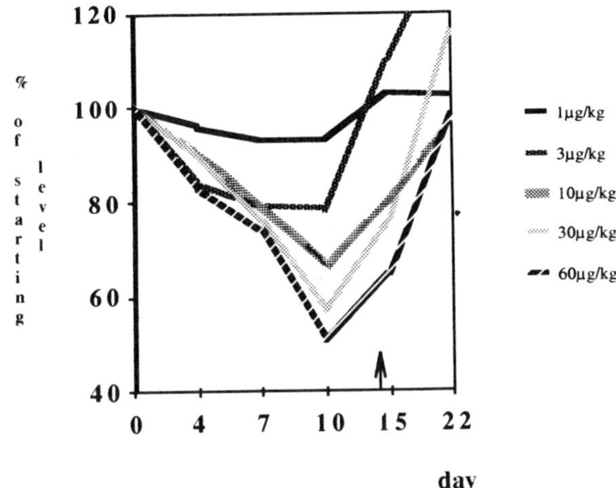

Figure 5. Course of platelet counts of three patients each
(median) treated at different dose levels of G-CSF as indicated, day 1-14.

The granulocytes induced by G-CSF in vivo were found to be
functionally normal or even activated as evidenced by measurement of oxygen radical release pre- and post administration of
G-CSF. However, as reported from in vitro data and animal
studies, the potential of G-CSF to functionally activate
phagocytes is almost neglegible when compared to GM-CSF. For
the clinical use however, it may be advantageous to just increase the number of competent cells that may be triggered to
exert their function at local sites of demand.

GM-CSF was the first myeloid GF entered into clinical
trials. It is directly involved in eosinophil-, neutrophil-
and monocyte differentiation and synergizes with other cytokines in erythro- and thrombopoiesis (Fig. 1). In vivo administration of GM-CSF induced a dose dependent increase of

neutrophils, eosinophils and monocytes (10). Absolute WBC
count and eosinophilia were more pronounced after c.i.v. as
compared to s.c. application, the latter being preferable in
the clinical setting particularly in outpatients (11). A
transient thrombocytopenia was seen similar as with G-CSF.

Interestingly, the number of peripheral blood CFU-GM increased at higher doses, providing the opportunity to improve
peripheral stem cell harvest in the setting of autologous bone
marrow transplantation (ABMT) (10). Since the maturation time
of GM-CSF induced peripheral progenitor cells is much shorter
as compared to bone marrow derived stem cells, combined transplantation of both types of progenitors leads to very rapid
and sustained engraftment that reduces the risk of ABMT dra-
matically (12).

One of the most promising indications for application of
CSFs may be the prevention of chemotherapy induced cytopenia,
that is currently studied at several institutions. We have
compared the duration of neutropenia and the incidence of
related complications after a 1. neutropenia inducing chemotherapy cycle without and an identical 2. cycle with s.c. GM-
CSF started 2 days after discontinuation of chemotherapy. Evaluation of 40 chemotherapy cycles in summary demonstrated a
shortening of the neutropenic phase by almost 50 %. Similarly
febrile episodes, the incidence of mucositis, needs of paren-
teral antibiotics and days in hospital were significantly reduced due to GM-CSF administration. Similar findings have also
been obtained in ABMT, thus arguing for the use of the GM-CSF
prior to the harvest as well as after transplantation (13).

Aside from stimulation of hematopoiesis GM-CSF also
serves as a potent inducer of various phagocyte functions including cytokine production by monocytes (14,15). To extend
these studies on PMN activation we investigated whether PMN,
sharing several properties with monocytes, might similarly be
able to produce cytokines upon GM-CSF stimulation. As reported
recently by our group (16-18), several cytokine transcripts
and the respective proteins were detected in GM-CSF induced
PMN and PMN conditioned media by means of Northern blotting
and biological assays (Fig. 6 and Table 1). TNF, however, was
not secreted, may be due to a more restricted control of gene
expression. These studies confirmed the synthetic potential of
PMN as reported for c-fos or plasminogen activator (19,20) and
they stress the characterization of GM-CSF as an inflammatory
mediator. GM-CSF probably works as a central hormone in host
defense, mediating the interplay of immune response, inflammation, and hematopoiesis by direct pathways as well as by indirect effects via stimulation of phagocytes (Fig. 7).

In summary we found the HGF to be highly effective in
vivo as predicted from in vitro studies. They are able to
improve states of hematopoietic deficiency may be in renal
anemia or in chemotherapy induced cytopenia. In addition to
their proliferation inducing capacity the pleiotropic T cell
factors IL-3 and GM-CSF share characteristics of physiologic
inflammatory mediators, a fact, that may help to explain the
scenario or side effects not observed with G-CSF or EPO.

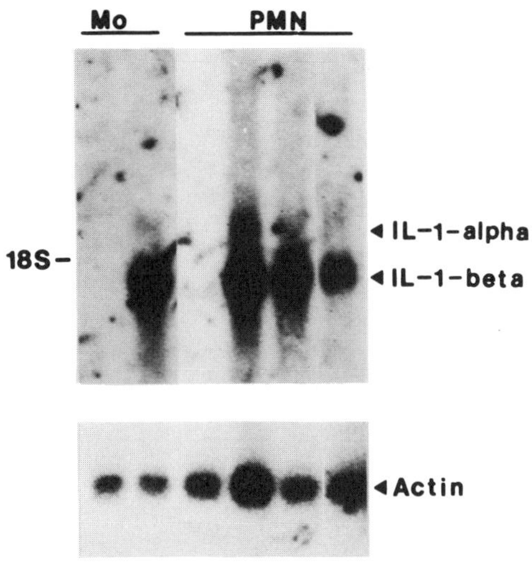

Figure 6. Accumulation of IL-1-beta and IL-1-alpha mRNA in GM-CSF (100 ng/ml) -induced polymorphonuclear neutrophils (PMN) after different time intervals as indicated. Monocytes (Mo) cultured in the presence of medium alone (ni) or LPS (10 ng/ml) + IFN-gamma (100 U/ml) (i) for 8 hours were used as control. Filters were rehybridized with an actin specific probe to ensure adequate RNA load of each lane.

Table 1. "Monokine" synthesis by GM-CSF induced PMN

	mRNA	protein
M-CSF	+	+
G-CSF	+	+
IL-1	+	+
TNF-alpha	+	-

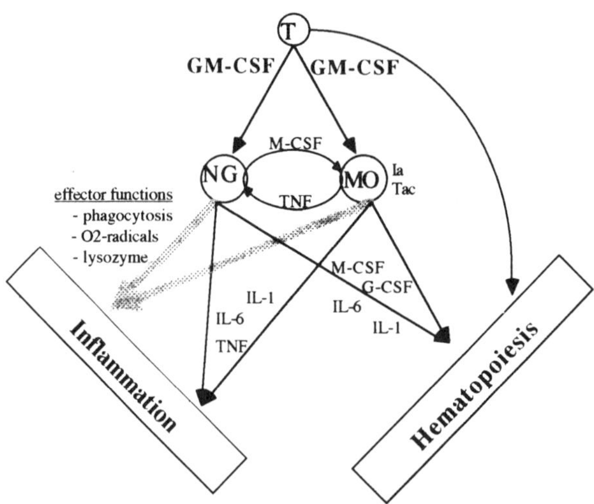

Figure 7. GM-CSF as a mediator of inflammatory and hematopoietic responses. The immunological system represented by the
activated T cell recruits the other components of the host defense system via GM-CSF secretion. Functional activation of
phagocytes as well as secondary mediators are involved in the
effector cascade.

REFERENCES

1. Herrmann, F., Mertelsmann, R.: Polypeptides controlling
    hematopoietic cell development and activation. <u>Blut</u> 58:
    117-128 (1989).
2. Eschbach, J.W., Egrie, J.C., Downing, M.R., Browne, J.K.,
    Adamson, J.W.: Correction of the anemia of end-stage
    renal disease with recombinant human erythropoietin. <u>N.
    Engl. J. Med.</u> 316:73-80 (1987).
3. Oster, W., Herrmann, F., Cicco, A., Gamm, H., Zeile, G.,
    Brune, T., Lindemann, A., Schulz, G., Mertelsmann, R.:
    Erythropoietin prevents chemotherapy-induced anemia: case
    report. <u>Blut</u> 60:88-92 (1990).
4. Oster, W., Herrmann, F., Gamm, H., Zeile, G., Lindemann,
    A., Brune, T., Kraemer, H.-P., Mertelsmann, R.: Erythro-
    poietin for the treatment of anemia of malignancy due to
    neoplastic bone marrow infiltration. <u>J. Clin. Oncol.</u>
    8:956-962 (1990).
5. Ihle, J.N., Keller, J., Oroszlan, S., Henderson, L.E.,
    Copeland, T.D., Fitch, F., Prystowsky, M.B., Goldwasser,
    E., Schrader, J.W., Palaszynsky, E., Dy, M., Lebel, B.:
    Biologic properties of homogenous interleukin 3. <u>J.
    Immunol.</u> 131:282-287 (1983).
6. Ganser, A., Lindemann, A., Seipelt, G., Ottmann, O.G.,
    Herrmann, F., Schulz, G., Mertelsmann, R., Hoelzer, D.:
    Effect of recombinant human interleukin-3 (rhIL-3) in
    patients with bone marrow failure - a phase I/II trial.
    <u>Blut</u> (Suppl. 1) 74:177 (1989) (abstract).
7. Lindemann, A., Herrmann, F., Ganser, A., Ottmann, O.,
    Hoelzer, D., Mertelsmann, R., Schulz, G.: Human recom-
    binant interleukin-3, a phase-I/II clinical study in:

Biologic properties of homogenous interleukin 3. <u>J. Immunol.</u> 131:282-287 (1983).

6. Ganser, A., Lindemann, A., Seipelt, G., Ottmann, O.G., Herrmann, F., Schulz, G., Mertelsmann, R., Hoelzer, D.: Effect of recombinant human interleukin-3 (rhIL-3) in patients with bone marrow failure - a phase I/II trial. <u>Blut</u> (Suppl. 1) 74:177 (1989) (abstract).

7. Lindemann, A., Herrmann, F., Ganser, A., Ottmann, O., Hoelzer, D., Mertelsmann, R., Schulz, G.: Human recombinant interleukin-3, a phase-I/II clinical study in: "Hematopoietic Growth Factors", R. Mertelsmann and F. Herrmann, eds., Marcel Dekker, New York, pp. 149-159 (1990).

8. Cannistra, S.A., Vellenga, E., Groshek, P., Rambaldi, A., Griffin, J.D.: Human granulocyte-macrophage colony-stimulating factor and interleukin-3 stimulate monocyte cytotoxicity through a tumor necrosis factor-dependent mechanism. <u>Blood</u> 71:672-676 (1988).

9. Lindemann, A., Herrmann, F., Oster, W., Haffner, G., Meyenburg, W., Souza, L.M., Mertelsmann, R.: Hematologic effects of recombinant human granulocyte colony-stimulating factor in patients with malignancy. <u>Blood</u> 74:2644-2651 (1989).

10. Herrmann, F., Schulz, G., Lindemann, A., Meyenburg, W., Oster, W., Krumwieh, D., Mertelsmann, R.: Hematopoietic responses in patients with advanced malignancy treated with recombinant human granulocyte-macrophage colony-stimulating factor. <u>J. Clin. Oncol.</u> 7:159-167 (1989).

11. Herrmann, F., Ganser, A., Lindemann, A., Wieser, M., Schulz, G., Hoelzer, D., Mertelsmann, R.: Stimulation of granulopoiesis in patients with malignancy by rhGM-CSF: assessement of two routes of administration. <u>J. Biol. Response Modif.</u> in press (1990).

12. Peters, W.P., Kurtzberg, I., Kirkpatrick, G., Atwater, S., Gilbert, C., Borowitz, M., Shpall, E., Jones, R., Ross, M., Affronti, M., Coniglio, D., Mathias, B., Oette, D.: GM-CSF primed peripheral blood progenitor cells coupled with autologous bone marrow transplantation will eliminate absolute neutropenia following high dose chemotherapy. <u>Blood</u> (Suppl. 1) 74:178 (1989) (abstract).

13. Herrmann, F., Schulz, G., Wieser, M., Kolbe, K., Nicolay, U., Noack, M., Lindemann, A., Mertelsmann, R.: Hematopoietic growth factors. Effect of granulocyte-macrophage colony-stimulating factor on neutropenia and related morbidity induced by myelotoxic chemotherapy. <u>Am. J. Med.</u> 89:619-624 (1990).

14. Weisbart, R.H., Golde, D.W., Clark, S.C., Wong, G.G., Gasson, J.C.: Human granulocyte-macrophage colony-stimulating factor is a neutrophil activator. <u>Nature</u> 314:361-363 (1985).

15. Lopez, A.F., Williamson, D.J., Gamble, J.R., Begley, C.G., Harlan, J.M., Klebanoff, S.J., Waltersdorph, A., Wong, G.G., Clark, S.C., Vadas, M.A.: Recombinant human granulocyte-macrophage colony-stimulating factor stimulates in vitro human neutrophil and eosinophil function, surface receptor expression and survival. <u>J. Clin. Invest</u> 78:1220-1228 (1986).

16. Lindemann, A., Riedel, D., Oster, W., Meuer, S.C., Blohm, D., Mertelsmann, R., Herrmann, F.: GM-CSF induces secretion of interleukin-1 by polymorphonuclear neutrophils. <u>J. Immunol.</u> 140:837-839 (1988).

17. Cicco, N.A., Lindemann, A., Content, J., Vandenbusche,
    J., Mertelsmann, R., Herrmann, F.: GM-CSF induction of
    interleukin-6 by human polymorphonuclear leukocytes.
    Blood 75:2049-2054 (1990).
18. Lindemann, A., Oster, W., Ziegler-Heitbrock, J.W.L.,
    Mertelsmann, R., Herrmann, F.: GM-CSF induces monokine
    secretion by polymorphonuclear leukocytes. J. Clin.
    Invest. 83:1308-1312 (1989).
19. Granelli-Piperno, A., Vasalli, J.-D., Reich, E.: Secre-
    tion of plasminogen activator by human polymorphonuclear
    leukocytes. J. Exp. Med. 146:1693-1706 (1977).
20. Heidorn, K., Kreipe, H., Radzun, H.J., Müller, R.,
    Parwaresch, M.R.: The proto-omcogene c-fos is transcrip-
    tionally active in normal human granulocytes. Blood
    70:456-459 (1987).

NEUTROPHIL FUNCTION IN POLYCYTHEMIA VERA

Jan Samuelsson

3rd Department of Internal Medicine
Södersjukhuset
S-10064 Stockholm, Sweden

INTRODUCTION

Polycythemia vera (PV) is a malignant myeloproliferative disorder characterized by the evolution of a single clone
of abnormally expanding pluripotent stem cells[1], resulting in
increased numbers of circulating erythrocytes, platelets and
granulocytes. Studies employing X-chromosome linked DNA probes
for the glucose-6-phosphate dehydrogenase[1] and phosphoglyce-
rate kinase[2] genes have shown that circulating polymorphonuclear granulocytes (PMN) in PV are derived from the malignant clone.

In 1972 Cooper et al. suggested that PMN were activated
in PV[3]. The basis for this concept were studies indicating increases in glycogen metabolism[4-6], phagocytic[3,7] and oxidative
metabolism[3]. However, as later research has shown, the only
incontestable results are those concerning glycogen metabolism. Thus an increased glycogen content[4], a 1.5 times greater
rate of glycogen turnover[5] and a mean increase of 33 % in PAS-
reactive material[6] in PV PMN has been reported. Studies of
phagocytosis and oxidative metabolism, on the other hand, have
yielded conflicting results. Enhanced phagocytosis of latex
particles[3] and yeast[7] has been found, whereas Corberand et
al.[8] using the same yeast assay found a phagocytic index that
wasn't elevated compared to controls. In the study by Cooper
et al.[3] an increased oxidative metabolism was implied by
higher oxygen consumption in both resting and latex particle
stimulated cells, as well as by increased nitroblue tetra-
zolium (NBT) reduction in resting PMN. The NBT results could
not be reproduced by Corberand et al.[8] who, instead, showed
normal NBT reduction in resting PMN and decreased NBT reduction after stimulation with latex particles. Work in our
laboratory has been aimed at resolving the existing contro-
versy by studying other neutrophil functions, all of which are
important markers of the degree of PMN activation. In contrast
to previous studies, in which unspecific stimuli (i. e.
phagocytosis) were used to elicit the functional response, we
have used stimuli with both defined as well as discrete pathways for activation of the neutrophil.

New Aspects of Human Polymorphonuclear Leukocytes
Edited by W.H. Hörl and P.J. Schollmeyer, Plenum Press, New York, 1991

PATIENTS AND METHODS

All patients had a PV diagnosis based on the Polycythemia Vera Study Group criteria[9] and only one patient received chemotherapy at the time of study. PMN were separated from heparinized blood using a one-step Percoll technique[10]. Luminol enhanced chemiluminescence (CL) was studied in a Chronolog Lumiaggregometer[11], superoxide anion production was analyzed using the SOD-inhibitable cytochrome C reduction method[12], and intracellular calcium concentrations were calculated from the change of Fura-2 fluorescence[13].

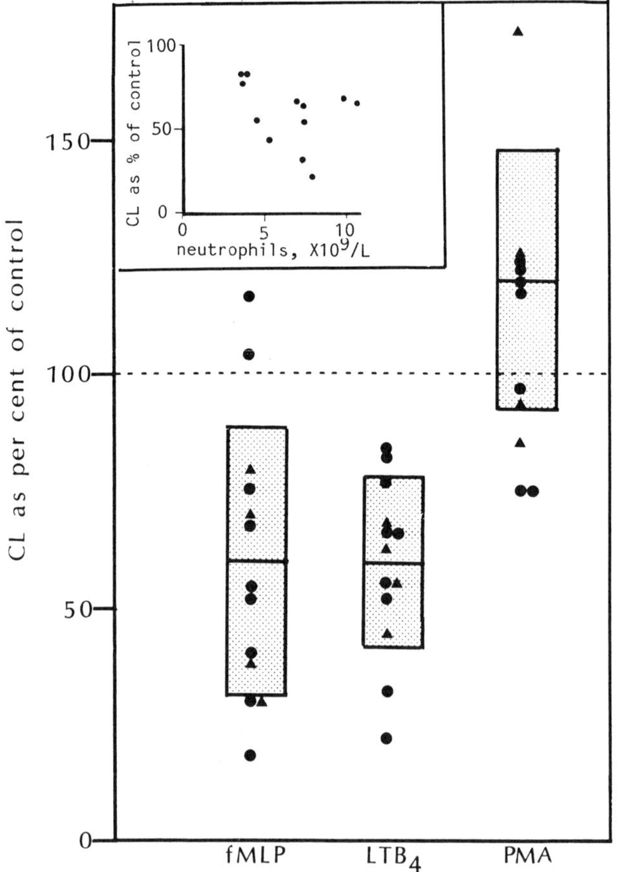

Figure 1. CL of PMN from PV patients, expressed as % of simultaneously run controls. The hatched areas show mean ± SD for the patients. indicates previous ^{32}P treatment. The insert shows LTB$_4$ induced CL related to peripheral blood neutrophil count in PV patients. Reproduced from Eur J Hematol[17].

Furthermore, chemotaxis was measured with a miniaturized Boyden chamber technique[14] and adherence was assessed in albumin-coated plastic dishes[15]. Finally, PMN production of leukotrienes B_4 and C_4 after stimulation with A23187 was analyzed using reversed phase HPLC technique[16]. All results are mean ± SD.

RESULTS

As shown in Figure 1, our first study[17] suggested the existence of a stimulus-specific defect in PV neutrophil oxidative metabolism, assessed by CL, after stimulation with surface receptor dependent stimuli. N-formyl-methionyl-leucyl-phenylalanine (fMLP) induced CL was 60 ± 29 % of control (p < 0.01) and leukotriene B_4 (LTB$_4$) induced CL was 59 ± 18 % (p < 0.01). In contrast, no impairment was detected after stimulation with phorbol myristate acetate (PMA), which stimulates protein kinase C directly[18].

No correlation could be found between fMLP- and LTB$_4$-induced CL and clinical parameters (phlebotomy frequency, disease duration, previous treatment) or peripheral blood counts. Also there was no difference in spontaneous CL or the kinetics of the stimulus-specific CL responses between patients and controls. Furthermore, spontaneous and fMLP induced adherence, as well as spontaneous migration and chemotactic responses to fMLP and LTB$_4$ (the latter depicted in Figure 2), were all found to be normal in PV.

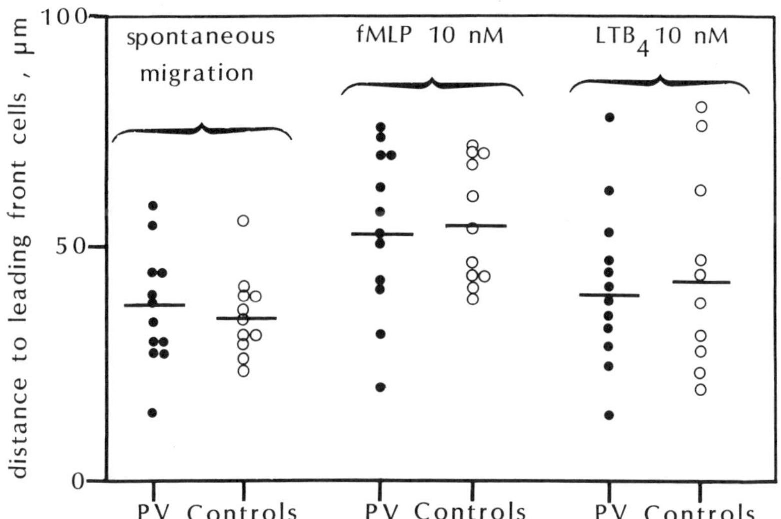

Figure 2. Migration of neutrophils in Boyden chambers. Values for patients ( ) and controls ( ) are given as net migration towards fMLP and LTB$_4$. Reproduced from Eur J Hematol[17].

It is well known that luminol enhanced CL is a complex reaction dependent on both superoxide anion (O_2^-) production[11] as well as substances from the myeloperoxidase (MPO) system[19,20]. The fact that PV neutrophils very seldom are MPO deficient[21], and the knowledge that PMN from patients with MPO deficiency exhibit virtually no CL in response to PMA[19], argued against MPO deficiency as an explanation for the impaired CL response that we had observed. In our next study we could also show that O_2^- production was reduced in PV neutrophils in response to fMLP, 0.3 ± 0.3 nmol $O_2^-/10^6$ PMN/min, compared with 0.6 ± 0.3 in controls ($p < 0.05$), whereas no difference in PMA induced O_2^- production was seen between patients and controls[22]. In simultaneously run experiments, PV neutrophils again showed reduced CL in response to FMLP (but not to PMA) and time to peak CL was the same as time to peak O_2^- production, proving that impaired CL reflected decreased O_2^- production.

Studies from our group have shown that an intact neutrophil LTB_4 forming capacity appears to be essential for the oxidative metabolism induced by fMLP in normal PMN[14]. It was therefore of interest to investigate if PV neutrophils showed any abnormalities in LTB^{14} production. However, no alterations in the production of LTB_4 and LTC_4 were observed in PV[23].

Which then is the exact step in the stimulus-response coupling answerable for the defective response to fMLP and LTB_4? This question remains unanswered at the present time. Both stimuli elicit a response by binding to specific surface receptors that exist in two affinity states. Since adherence and chemotaxis are thought to be mediated by binding of fMLP or LTB_4 to high affinity receptors[24,25], whereas PMN oxidative metabolism mainly depends on binding of the stimulus to low affinity receptors[24,25], it is conceivable that an altered function and/or number of low affinity receptors on PV neutrophils is responsible for our results. Support for this theory is derived from the fact that sodium fluoride, which acts at the level of the regulatory G protein[26], confered a normal CL response in PV PMN with simultaneously decreased CL in response to fMLP[22]. On the other hand, we have also obtained data that somewhat argues against this hypothesis. In PV PMN exhibiting reduced CL in response to fMLP, fMLP-elicited cytosolic calcium changes were normal[22]. This suggests that binding of fMLP to PV neutrophils generate a normal signal that is transmitted into the cell resulting in the synthesis of inositol-1,4,5-triphosphate which mediates the release of intracellular calcium [27]. It should be borne in mind, however, that binding of chemoattractant to either high or low affinity receptors is sufficient to initiate a calcium response[28,29]. Also, chemotactic levels of fMLP have been reported to generate an optimal calcium response without triggering superoxide anion production[30].

If we assume that fMLP receptor function, which currently is being studied using ^{3}H-fMLP binding assay, is found to be normal in PV PMN, what alternative explanation could there be for the observed defect? One possible option could be an impairment in the generation of diacylglycerol (DAG). It has recently been suggested that products resulting from inositol-phospholipid hydrolysis, if sufficiently sustained, triggers

an influx of calcium resulting in a prolonged rise in intracellular calcium. This leads to activation of a second
source of DAG, perhaps from hydrolysis of phosphorylcholine[31],
that is necessary for activation of the respiratory burst[32].
It is thus possible that impaired DAG generation could account
for the decreased superoxide anion production in PV PMN that
we have described and this possibility is being evaluated
seperately.

Conclusions

We have found no data to support the view that PMN in PV
are activated in a clear cut way. The finding of normal
spontaneous adherence and migration argues strongly against
this concept. We have instead provided evidence for a stimulus-specific defect in the oxidative metabolism of PV PMN
which is attributable to reduced superoxide anion production.
The challenge is now to explore the possibility that this
defect might reflect changes that are also relevant in the
malignant stem cell. Therefore, further biochemical and
genetic studies are under way in the hope of obtaining
knowledge that might cast new light on the underlying mechanisms behind the development of PV.

REFERENCES

1. Adamson, J.W., Fialkow, P.J., Murphy, S., Prchal, J.F.,
    Steinmann, B.A.: Polycythemia vera: stem-cell and
    probable clonal origin of the disease. <u>N. Engl. J. Med.</u>
    295:913 (1976).
2. Gilliland, G., Levy, J., Perrin, S., Blanchard, K., Bunn,
    H.F.: Determination of clonality in myeloproliferative
    diseases using polymerase chain reaction. <u>Clin. Res.</u>
    37(2):601 (1989).
3. Cooper, M.R. DeChatelet, L.R., McCall, C.E., Spurr, C.L.:
    The activated phagocyte of polycythemia vera. <u>Blood</u>
    40:366 (1972).
4. Wagner, R.: Studies on the physiology of white blood
    cells. The glycogen content of leukocytes in leukemia and
    polycythemia. <u>Blood</u> 2:235 (1947).
5. Luganova, I.S., Seitz, I.E.: Glycogen content and normal
    metabolism in normal and leukemic human leukocytes. <u>Fed.
    Proc. Transpl. Suppl.</u> 22:1058 (1963).
6. Gahrton, G.: The periodic acid-Schiff reaction in neutro-
    phil leukocytes in chronic myeloproliferative diseases. A
    micro-spectrophotometric study. <u>Scand. J. Hematol.</u> 3:106
    (1966).
7. Brandt, L.: Studies on the phagocytic activity of poly-
    cythemia vera leukocytes. <u>Scand. J. Hematol. Suppl.</u> 2:54
    (1967).
8. Corberand, J., Laharrague, P., De Larrad, B., Ngyen, F.,
    Pris, J.: Phagocytosis in myeloproliferative disorders.
    <u>Am. J. Clin. Pathol.</u> 74:301 (1980).
9. Berlin, N.I.: Diagnosis and classification of the poly-
    cythemias. <u>Semin. Hematol.</u> 12(4):339 (1976).
10. Ringertz, B., Palmblad, J. Rådmark, O., Malmsten, C.L.:
    Leukotriene induced neutrophil aggregation in vitro. <u>FEBS
    Lett.</u> 147:180 (1982).

11. Palmblad, J., Gyllenhammar, H., Lindgren, J.Å., Malmsten, C.L.: Effects of f-Met-Leu-Phe on oxidative metabolism of neutrophils and eosinophils. J. Immunol. 132:3041 (1984).

12. Goldstein, I.M., Roos, D., Kaplan, H.B., Weissmann, G.: Complement and immunoglobulins stimulate superoxide production by human leukocytes independently of phagocytosis. J. Clin. Invest. 56:1155 (1975).

13. Grynkiewitz, G, Poenie, M., Tsien, R.Y.: A new generation of Ca^{2+} indicators with greatly improved fluorescence proper-ties. J. Biol. Chem. 260:3440 (1985).

14. Gyllenhammar, H., Palmblad, J., Ringertz, B., Hafström, I., Borgeat, P.: Rat neutrophil function and leukotriene generation in essential fatty acid deficiency. Lipids 23:89 (1988).

15. Lindström, P., Palmblad, J.: Dietary linoleic acid and rat neutrophil adhesion and chemotaxis. J. Clin. Lab. Immmunol. 25:77 (1988).

16. Miyamoto, T., Lindgren, J.Å., Samuelsson, B.: Isolation and identification of lipooxygenase products from the rat central nervous system. Biochim. Biophys. Acta 922:372 (1987).

17. Samuelsson, J., Lindström, P., Palmblad, J.: Stimulus-specific defect in oxidative metabolism of polymorphonuclear granulocytes in polycythemia vera. Eur. J. Hematol. 41:454 (1988).

18. Castagna, M., Tahai, Y., Kaibuchi, K., Sano, K., Kikhawa, K., Nishizuka, Y.: Direct activation of calcium-activated, phospholipid-dependent protein kinase by tumor-promoting phorbol ester. J. Biol Chem. 257:7847 (1982).

19. DeChatelet, L.R., Long, G.D., Shirley, P.S., Bass, D.A., Thomas, M.J., Henderson, F.W., Cohen, M.S.: Mechanisms of the luminol-dependent chemiluminescence of human neutrophils. J. Immunol. 129:1589 (1982).

20. Dahlgren, C., Stendahl, O.: Role of myeloperoxidase in luminol dependent chemiluminescence of polymorphonuclear. Infect. Immunol. 39:736 (1983).

21. Bendix-Hansen, K.: Myeloperoxidase-deficient polymorphonuclear leukocytes (VII): incidence in untreated myeloproliferative disorders. Scand. J. Hematol. 36:8 (1986).

22. Samuelsson, J., Berg, A.: Studies on a stimulus-specific defect in the oxidative metabolism of neutrophils in polycythemia vera: relation to intracellular calcium concentrations and priming with granulocyte-macrophage colony-stimulating factor. Eur. J. Clin. Invest. (submitted).

23. Stenke, L., Samuelsson, J., Palmblad, J., Dabrowski, L., Reizenstein, P., Lindgren, L.Å.: Elevated white blood cell synthesis of leukotriene C_4 in chronic myelogenous leukemia but not in polycythemia vera. Br. J. Hematol. (submitted).

24. Snydermann, R., Pike, M.C.: Chemoattractant receptors on phagocytic cells. Ann. Rev. Immunol. 2:257 (1984).

25. Goldmann, D.W., Goetzl, E.J.: Heterogeneity of human polymorphonuclear leukocyte receptors for leukotriene B_4. Identification of a subset of high affinity receptors that transduce the chemotactic response. J. Exp. Med. 159:1027 (1984).

26. Strnad, C.F., Parente, J.E., Wong, K.: Use of fluoride ion as a probe for the guanine nucleotide-binding protein

involved in the phosphoinositide-dependent neutrophil
transduction pathway. <u>FEBS Lett.</u> 206:20 (1986).
27. Prentki, M., Wollheim, C.B., Lew, P.D.: Ca^{2+} homostasis
in permeabilized human neutrophils. <u>J. Biol. Chem.</u>
259:13777 (1984).
28. Goldmann, D.W., Gifford, L.A., Olson, D.M., Goetzl, E.J.:
Transduction by leukotriene B_4 receptors of increases in
cytosolic calcium in human polymorphonuclear leukocytes.
<u>J. Immunol.</u> 135:525 (1985).
29. Seligmann, B.E., Fletcher, M.P., Gallin, J.I.: Adaption
of human neutrophil responsiveness to the chemoattractant
n-formyl-methionyl-leucyl-phenyl-alanine. <u>J. Biol. Chem.</u>
257:6280 (1982).
30. Korchak, H.M., Vienne, K., Rutherford, L.E., Wilkenfeld,
C., Finkelstein, M.C., Weissmann, G.: II. Temporal
analysis of changes in cytosolic calcium and calcium
efflux. <u>J. Biol. Chem.</u> 259:4076 (1984).
31. Truett III, A.P., Snydermann, R., Murray, J.J.: Stimulation of phosphorylcholine turnover and diacylglycerol
production in human polymorphonuclear leukocytes. Novel
assay for phosphorylcholine. <u>Biochem. J.</u> 260:909 (1989).
32. Truett III, A.P., Verghese, M.W., Dillon, S., Snyderman,
R.: Calcium influx stimulates a second pathway for
sustained diacylglycerol production in leukocytes
activated by chemoattractants. <u>Proc. Natl. Acad. Sci. USA</u>
85:1549 (1988).

HUMAN TUMOR CELL UROKINASE-TYPE PLASMINOGEN ACTIVATOR (uPA): DEGRADATION OF THE PROENZYME FORM (pro-uPA) BY GRANULOCYTE ELASTASE PREVENTS SUBSEQUENT ACTIVATION BY PLASMIN

Manfred Schmitt, Naohiro Kanayama[*], Fritz Jänicke, Reimar Hafter, and Henner Graeff

Frauenklinik der Technischen Universität München, Klinikum rechts der Isar, Ismaningerstr. 22, D-8000 München, FRG

Hamamatsu Medical School, Hamamatsu University, Japan[*]

SUMMARY

When human granulocytes were stimulated with the chemotactic peptide FNLPNTL (N-formyl-norleucyl-leucyl-phenylalanyl-norleucyl-tyrosinyl-lysin; in the presence of cytochalasin B), proteolytic enzymes were released which prevented activation of tumor-cell derived pro-uPA by plasmin. Elastase was identified by use of eglin C (elastase inhibitor) and an inhibitory monoclonal antibody to elastase as the functional proteolytic enzyme in these granulocyte supernatants.

Purified human granulocyte elastase cleaves pro-uPA at amino acid position Ile^{159}-Ile^{160} thus generating an enzymatically inactive two-chain form of uPA, as judged by N-terminal amino acid sequence analysis. An additional minor elastase-mediated cleavage site was detected at position Thr^{165}-Thr^{166}. This form of uPA was indistinguishable by SDS-PAGE from plasmin-generated enzymatically active HMW-uPA. Action of plasmin on the proenzyme form of uPA (pro-uPA) generates an enzymatically active uPA-molecule (high molecular weight form; HMW-uPA) which is cleaved at amino acid position Lys^{158}-Ile^{159} (M_r=33,000 (B-chain) and 22,000 (A-chain). Thus elastase cannot substitute for plasmin in the proteolytic activation of pro-uPA to enzymatically active HMW-uPA. Enzymatically active HMW-uPA, however, was not affected by elastase. Elastase-containing granulocytes were identified by immunohistochemical staining of elastase in breast cancer tissue. Granulocytes were located close to the tumor cells and also in the tumor stroma surrounding the tumor nests. These tumor cells contain pro-uPA. Evidently, the conversion of tumor cell pro-uPA into enzymatically active HMW-uPA is controled by elastase released from granulocytes into the tumor tissue.

INTRODUCTION

Plasminogen activators are major mediators of extracellular proteolysis in benign as well as malignant conditions. In tumor cells of breast, prostate, cervix, and colon cancer tissue elevated levels of the urokinase-type plasminogen activator (uPA) are produced and secreted as an enzymatically inactive proenzyme form (pro-uPA) **(1-3)**. This increase of pro-uPA/uPA in tumor cells <u>in vivo</u> has been associated with increased tumor growth and metastatic potential **(4)**. Pro-uPA may be converted by small amounts of plasmin into the enzymatically active high-molecular-weight form of uPA (HMW-uPA) which subsequently converts plasminogen into the serine protease plasmin **(1,2)**. HMW-uPA is a serine protease composed of two polypeptide chains (M_r=22,000 and 33,000) connected by disulfide bridges **(2,3)**. Both, pro-uPA and HMW-uPA, may bind with high affinity via amino acid sequence 20-30 to specific receptors on tumor cells **(2)**. The naturally occuring plasminogen activator inhibitors (PAI-1/2) may inactivate soluble or receptor-bound HMW-uPA. PAI-1/2 do not bind to pro-uPA **(5).** The protease thrombin, however, is known to degrade pro-uPA into an enzymatically inactive uPA-molecule which cannot be activated by plasmin, anymore. Here we report that elastase released from activated human granulocytes also degrades and inactivates tumor cell pro-uPA .

MATERIALS AND METHODS

MATERIALS

Reagents were obtained from the sources indicated in parantheses. Pro-uPA (specific activity of 135,000 units / mg) produced by the kidney tumor cell line TCL 598 (Sandoz, Nürnberg, FRG). HMW-uPA (Serono, Freiburg, FRG). Human granulocyte elastase (Protogen, Läufelfingen, Switzerland). Human plasmin and fibrinogen (KabiVitrum, München). Prestained gel standards (Pharmacia, Freiburg, FRG). Aprotinin and test-thrombin (Behring-Werke, Marburg, FRG). Bovine serum albumin (BSA), cytochalasin B and phosphatase-anti-phosphatase complex (PAP) (Sigma, München, FRG). N-formyl-Nle-Leu-Phe-Nle-Tyr-Lys (FNLPNTL) and substrates S-2444 and methoxysuccinyl-Ala-Ala-Pro-Val-paranitroanilide (Bachem, Hannover, FRG). Mouse anti-rabbit-IgG, mouse monoclonal antibody to human granulocyte elastase (clone M752), and alkaline phosphatase-anti-alkaline-phosphatase complex (APAAP) (Dakopatts, Hamburg, FRG). Monoclonal antibody #377 and #394 to human uPA (American Diagnostica, New York, USA). Peroxidase-conjugated avidin (Dianova, Hamburg, FRG). 3,3',5,5'-tetramethylbenzidine (TMB) (Serva, Heidelberg, FRG). Rabbit-anti-human granulocyte elastase was provided by Dr. Terao (Hamamatsu, Japan) **(8)**. Eglin C was donated by Dr. H. Fritz, Ludwig-Maximilians-University, München, FRG, and inhibitory murine monoclonal antibody to human granulocyte elastase by Dr. T. Cotter, Maynooth, Ireland. All other reagents were of analytical grade.

METHODS

Immunohistochemical localization of elastase or uPA in breast cancer tissue sections

Specimens of patients with breast carcinoma were fixed in buffered formalin and then embedded in paraffin. Sections were cut and mounted on microscope slides. These paraffin-embedded sections were dewaxed in Roti-Histol (a xylene substitute), followed by isopropanol and 96 % (v/v) ethanol, rehydrated in PBS and then processed for the presence of pro--uPA/uPA and also for elastase by applying specific monoclonal antibodies to uPA (#394) or to elastase (M752) in combination with the APAAP--staining technique essentially as described (9).

SDS-PAGE

SDS-PAGE was performed according to Laemmli (10) applying 15 % acrylamide gels. Prestained proteins served as molecular weight markers. Samples were reduced by 10 mM 2-mercaptoethanol. (10 min, 100 °C). Proteins were stained with Coomassie Blue R-250 (Serva, Heidelberg, FRG) and the gels destained with 7 % acetic acid.

Determination of elastase by ELISA

The wells of a microtiter plate were coated with 100 µl of moAB M752 to elastase (0.4 µg / ml of 50 mM bicarbonate buffer, pH 9.6; 4 °C, 16 h) and then blocked with TBS-2% BSA (1 h, 37 °C). 100 µl of TBS-2 % BSA containing various amounts of granulocyte supernatants were added per well and the plate incubated (1 h, 23 °C). After washing in PBS, 100 µl of rabbit-IgG-anti-elastase (4 µg / ml TBS-0.05 % Tween 20) were added (1 h, 23 °C). Subsequently, after additional washing steps, mouse-IgG-anti-rabbit-IgG was added to the wells (1 h, 23 °C). After still another washing step, rabbit-PAP was added (1 h, 23 °C). After the final washing steps, 100 µl of substrate solution was added (0.1 M sodium acetate-citrate buffer, pH 6.0, containing 42 nM TMB and 0.0045 % H_2O_2). The reaction was terminated by addition of 50 µl of 1 M H_2SO_4 / well and optical density read at 450 nM with a computer-interfaced ELISA reader (Titertek Multiscan; Flow Laboratories, Meckenheim, FRG).

Elastase activity

The enzymatic activity of elastase was measured with 0.67 mM of the chromogenic amidolytic substrate methoxysuccinyl-Ala-Ala-Pro-Val-paranitroanilide in 0.1M Tris-HCl, pH 8.3, containing 960 mM NaCl. Enzymatic activity was measured by a change in absorbance at 405 nm. Purified granulocyte elastase (0 - 0.25 U / ml) served as standard.

<u>Determination of pro-uPA /uPA by ELISA or enzymatic activity</u>

Pro-uPA/uPA-antigen was determined by ELISA as described **(11)**. Briefly, wells of a microtiter plate were coated with 100 µl of a solution containing 1.5 µg moAB #394 / ml of 50 mM bicarbonate buffer, pH 9.6 (16 h, 4 °C) and then blocked with TBS-2 % BSA (1 h, 37 °C). After sample addition (100 µl in TBS-2 % BSA-0.05 % Tween 20, 16 h, 4 °C) biotinylated moAB #377 to uPA was added (1 h, 23 °C). After washing, avidin-peroxidase in 20 mM Tris-HCl-0.5 M NaCl-0.05 % Tween 20, pH 8.5, was added (1 h, 23 °C). Peroxidase activity was detected with the same substrate solution as described for the elastase-ELISA. The uPA-ELISA detects pro-uPA and the proteolytically degraded forms of uPA, as well **(12)**. Detection limit: 40 pg pro-uPA/uPA per ml.

Amidolytic activity of HMW-uPA was measured in 0.1 M Tris-HCl, pH 8.8, containing 0.1 M NaCl and 0.01 % BSA using the synthetic substrate S-2444 **(13)**. Fibrin clot lysis assay was performed according to Kruithof et al. **(14)** using fibrin gels which contain excess of plasminogen.

<u>Purification of human peripheral blood granulocytes</u>

Granulocytes were purified according to Boyum **(15)**. Viability was >98 % as judged by trypan blue exclusion. Contaminating erythrocytes were lysed by short exposure (5 min) of cells to lysis buffer (160 mM NH_4Cl, 12 mM $NaHCO_3$, 0.1 mM EDTA, pH 7.3). Cells were stored at 23 °C in PBS-2 % BSA without calcium and magnesium salts.

<u>Release of proteases (degranulation) by granulocytes stimulated with the chemotactic peptide FNLPNTL in the presence of cytochalasin B</u>

5×10^6 granulocytes in 1 ml PBS (containing 0.5 mM $MgCl_2$, 1 mM $CaCl_2$, 0.1 % BSA) were incubated with 5 µg cytochalasin B (5 min, 37 °C). 10 µl of 10^{-6} M FNLPNTL were added (30 min, 37 °C) and supernatants collected by low speed centrifugation (250 x g, 5 min). The supernatants were kept frozen at - 80 °C in the presence of 10 mM phenylmethansulfonylfluoride.

<u>Identification of proteolytic cleavage site in pro-uPA by sequence analysis</u>

370 µg of pro-uPA in 1 ml PBS were digested with 20 µg (0.5 units) of purified granulocyte elastase in 200 µl of 62 mM Tris-HCl, pH 7.5 (30 min, 37 °C). Digestion of single-chain pro-uPA into two-chain uPA was verified by SDS-PAGE. The digested sample and an undigested control sample were dialyzed against 3 % acetic acid. An aliquot of each sample was sequenced by Edman degradation in a prototype spinning-cup sequenator **(16)**. The amino acid derivatives were identified by isocratic HPLC analysis.

RESULTS

Degradation of tumor cell pro-uPA by proteases

The enzymatically inactive single-chain pro-uPA, may be converted by proteases into the two-chain form of uPA, HMW-uPA **(1-3, 6,7,17)**. To test the effect of proteases on tumor-cell derived pro-uPA <u>in vitro</u>, four different enzymes (granulocyte elastase, thrombin, plasmin, trypsin) were incubated with pro-uPA (purified from tissue culture supernatants of the human kidney tumor cell line TCL 598) and then the samples subjected to SDS-PAGE. All these enzymes degraded pro-uPA (M_r=55,000) into a molecule consisting of two major polypeptide chains of M_r=33,000 (B-chain) and M_r=22,000 (A-chain) connected by disulfide bonds **(Fig. 1)**.

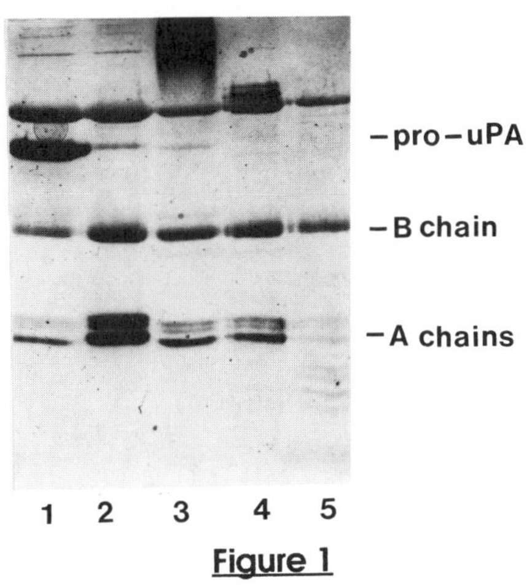

<u>Figure 1</u>

SDS-PAGE analysis (15 % acrylamide, reducing conditions) of pro-uPA treated with proteases

5 µg of pro-uPA in 30 µl TBS were treated with elastase, thrombin, plasmin, or trypsin (30 min, 37 °C).

Lane 1:   Pro-uPA
Lane 2:   Pro-uPA treated with 0.45 µg elastase   (500 nM)
Lane 3:   Pro-uPA treated with 0.73 µg thrombin (500 nM)
Lane 4:   Pro-uPA treated with 0.5 µg plasmin   (200 nM)
Lane 5:   Pro-uPA treated with 0.3 µg trypsin   (500 nM)

<u>Note:</u>   The action of elastase, thrombin, plasmin, or trypsin on pro-uPA produced polypeptide chains of similar molecular mass.
(A-chain(s): M_r=22,000; B-chain: M_r=33,000).

The prominent stained polypeptide band above pro-uPA represents albumin which was added by the manufacturer to the pro-uPA preparation as a stabilizing agent.

As also shown by SDS-PAGE, enzymatically active HMW-uPA (prepared from purified tumor cell pro-uPA by plasmin treatment or HMW-uPA purified from urine) was not degraded by these proteases **(Fig. 2)**. When HMW-uPA was treated with the same concentrations of plasmin, trypsin, thrombin or elastase as described for pro-uPA, no significant degradation of the polypeptide chains was observed. At high concentration of elastase (1 μM), however, significant cleavage of the A-chain and the B-chain into several polypeptide chains was observed (data not shown).

The elastase cleavage site in pro-uPA was evaluated by N-terminal amino acid sequence determination of the first 16 amino acids **(17)**. The elastase-treated pro-uPA was analyzed in parallel with untreated pro-uPA. In the undigested reference the known sequence of pro-uPA (first 16 amino acids) was found. In the elastase-treated sample two cleavage sites were found. The prominent cleavage site is between Ile^{159} and Ile^{160}, the second but minor one between Thr^{165} and Thr^{166} (approximate ratio of 5:1). The new sequences start with Ile^{160} and Thr^{166}, respectively. This is one (position 159 / 160) and seven (position 165 / 166) residues behind the begin of the plasmin-derived B-chain.

Degradation of pro-uPA by thrombin or granulocyte elastase prevents subsequent activation by plasmin

Treatment of pro-uPA with plasmin or trypsin generated enzymatically active uPA (HMW-uPA). In contrast, exposure of pro-uPA to thrombin or purified granulocyte elastase did not result in an enzymatically active uPA **(Fig. 3)**. When these samples were assessed for their capacity to split the substrate S-2444 **(upper graph)** or lyse fibrin gels in the presence of plasmin(ogen) **(lower graph)**, only the pro-uPA samples treated with plasmin or trypsin were enzymatically active. Hence, thrombin or elastase cannot substitute for plasmin in pro-uPA activation <u>in vitro</u>.

The proteolytic treatment of pro-uPA by elastase or thrombin not only generate an enzymatically inactive two-chain uPA-molecule but also impaired pro-uPA in such a way that subsequent activation of these uPA-molecules by plasmin was not possible anymore. Thrombin- or elastase-treated samples which were subjected to an additional plasmin treatment were also tested for their amidolytic or fibrinolytic activity. No enzymatic activity was generated when pro-uPA was treated with elastase or thrombin, first, and then subjected to plasmin treatment. The dose-dependent inactivation for elastase is shown in **Fig. 4**.

The data obtained with the enzymatically inactive thrombin-treated pro-uPA confirmed the observations made by Gurewich and Panell **(6)** and Ichinose et al. **(7)**. The new finding is that elastase treatment of pro-uPA also prevented the conversion of pro-uPA by the subsequent addition of plasmin **(17)**. The ability of pro-uPA to be converted by plasmin into an enzymatically active uPA-molecule decreased with increasing elastase concentration prior to plasmin treatment. Complete inhibition was observed at 100 nM elastase. Even at a concentration as low as 0.5 nM elastase, a loss of 35 % in activity was observed. When pro-uPA was incu-

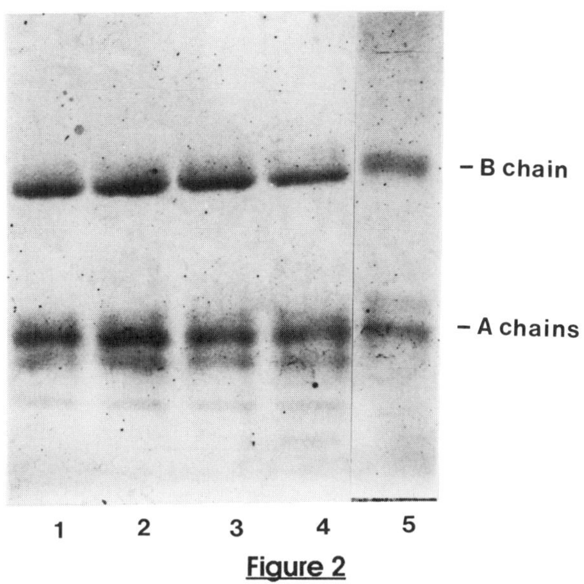

Figure 2

SDS-PAGE analysis (15 % acrylamide, reducing conditions) of HMW-uPA treated with proteases

5 µg of HMW-uPA (Serono) in 30 µl TBS were treated with elastase, thrombin, plasmin, or trypsin (30 min, 37 °C).

Lane 1: HMW-uPA
Lane 2: HMW-uPA treated with 0.45 µg elastase (500 nM)
Lane 3: HMW-uPA treated with 0.73 µg thrombin (500 nM)
Lane 4: HMW-uPA treated with 0.5 µg plasmin (200 nM)
Lane 5: HMW-uPA treated with 0.3 µg trypsin (500 nM)

Note: The positions of the A- and B-chain of HMW-uPA in the gel were not affected by the protease treatment.

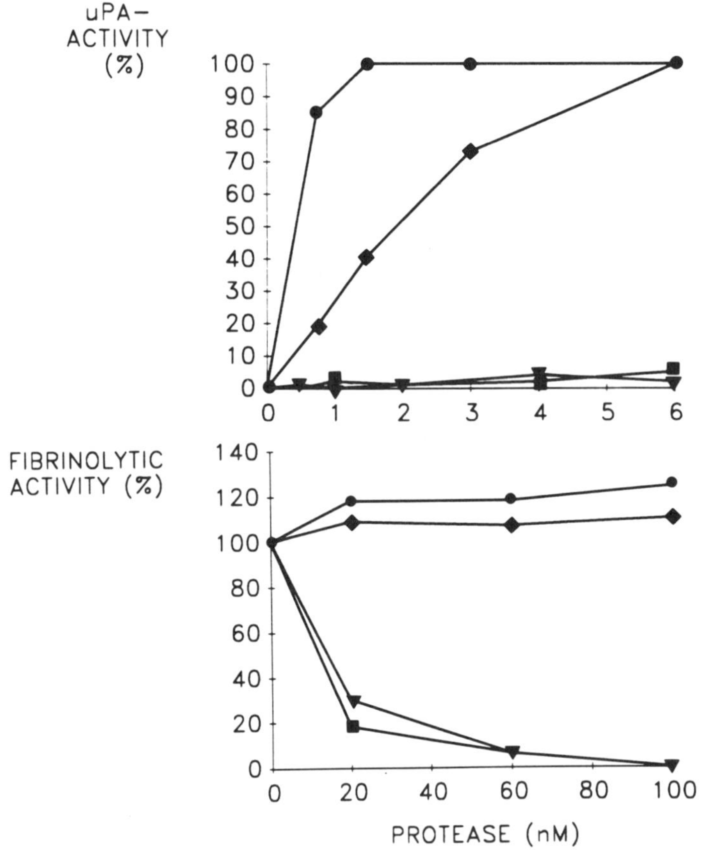

Figure 3

Activation of pro-uPA by proteases

50 units pro-uPA in 50 µl of 50 mM Tris-HCl, pH 7.3, were incubated with plasmin, trypsin, elastase, or thrombin with the indicated concentrations (0-6 nM) for 30 min at 37 °C. uPA activity was assessed by amidolytic assay applying substrate S-2444 (upper graph) and by fibrin clot lysis (lower graph). uPA amidolytic activity generated by 50 units of pro-uPA (which had been treated with 6 nM plasmin) was defined as 100 %. Fibrin lysis volume produced by 50 units of pro-uPA / 50 µl Tris-HCl, pH 7.3, was defined as 100 % of fibrinolytic activity.

Plasmin (●), trypsin (◆), elastase (■), thrombin (▼).

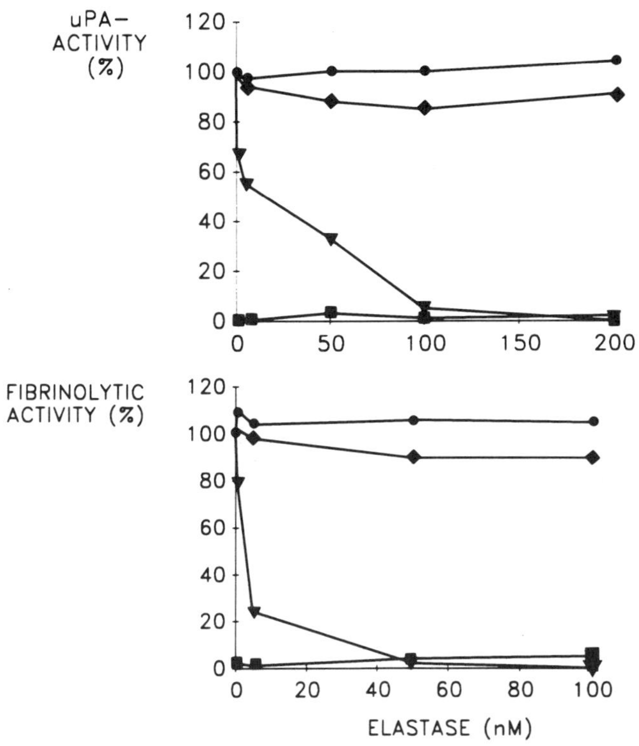

<u>Figure 4</u>

Effect of elastase on pro-uPA activation

Upper graph. For determination of **amidolytic activity,** pro-uPA (50 units in 50 mM Tris-HCI, pH 7.3) was incubated for 30 min, at 37 °C in the presence or absence of proteases. Pro-uPA treated with plasmin in the absence of elastase is defined as 100 % of activity.

1) Pro-uPA treated with elastase, first (30 min, 37 °C) and then 2.8 nM plasmin added (45 min, 37 °C) (▼).
2) Pro-uPA incubated with 2.8 nM plasmin, first (45 min, 37 °C), and then elastase added (30 min, 37 °C) (●).
3) Pro-uPA treated with elastase, first (30 min, 37 °C), in the presence of 50 % plasma, then addition of plasmin (45 min, 37 °C) (◆).
4) Elastase, only (75 min, 37 °C) (■).

Lower graph. For determination of **fibrinolytic activity,** pro-uPA (50 units in 50 mM Tris-HCI, pH 7.3) was layered over fibrin gels (containing trace amounts of plasmin and sufficient plasminogen). Lysis caused within 2 h at 37 °C is defined as 100 % activity.

1) Pro-uPA incubated with elastase (30 min, 37 °C) (▼).
2) Pro-uPA incubated with plasmin, first (45 min, 37 °C), and then elastase added (30 min, 37 °C) (●).
3) Pro-uPA treated with elastase in the presence of 50 % plasma (30 min, 37 °C) (45 min, 37 °C) (◆).
4) Elastase, only (75 min, 37 °C) (■).

bated with plasmin, first, and then elastase (0 - 200 nM) added, no loss of pro-uPA activity was detected. Under these conditions, very little fibrinolytic activity was determined in aliquots containing elastase, only.

Identification of elastase as the functional protease in granulocyte supernatants

Supernatants of human granulocytes, which had been stimulated with FNLPNTL, contained proteases which prevented conversion of tumor cell pro-uPA to enzymatically active HMW-uPA in comparison to supernatants obtained from unstimulated cells **(Table 1)**. In the presence of 50 % plasma, however, only slight inactivation of pro-uPA by elastase was observed **(Fig. 5)**. Human granulocytes also produce pro-uPA which is stored in specific granules **(18)**. Under the conditions applied, however, (short-term stimulation with chemotactic peptide FNLPNTL in the presence of cytochalasin B) no significant release of pro-uPA/uPA into cell supernatants of stimulated granulocytes was detected **(Table 2)**.

The proteolytic activity in supernatants of FNLPNTL-stimulated granulocytes was identified as elastase by incubation with either eglin C (elastase inhibitor) or a moAB to elastase (blocks activity) **(Table 1)**. Supernatants from granulocytes stimulated with FNLPNTL in the presence of cytochalasin B were incubated with eglin C and then mixed with pro-uPA. Supernatants were also incubated with moAB to elastase and then mixed with pro-uPA in PBS (30 min, 37 °C). Fibrin clot lysis and amidolytic activity were determined. In the presence of 12 µM eglin C or 15 µM of moAB to elastase the capability of pro-uPA to be converted into active HMW-uPA by subsequent treatment with plasmin was preserved. Thus by these inhibition experiments the proteolytic activity in supernatants of stimulated granulocytes was identified as elastase **(17)**.

Localization of elastase and uPA in breast cancer tissues

Cells containing uPA or elastase were localized in formalin-fixed paraffin-embedded human breast cancer tissue by moAB to the antigens. uPA was localized in the cytoplasm and on the plasma membrane of tumor cells **(Fig. 6)**. Elastase-containing cells were localized in the tumor tissue and also in the tumor stroma surrounding the tumor nests **(Fig. 7)**. These cells display polymorph-shaped nuclei and most probably represent granulocytes. The staining pattern of some of these phagocytic cells was irregular indicating release of elastase into the tumor stroma. Breast cancer cells did not stain for elastase.

20 µg / ml of moAB to uPA (#394) or elastase (M752) in PBS-1% BSA were added to breast cancer tissue sections (16 h, 4 °C). After washing in PBS, the tissue sections were reacted with 20 µg / ml of Ig-rabbit anti-mouse Ig in PBS-2 % BSA (1 h, 23 °C), washed with PBS and then a 1:50 dilution of mouse-APAAP in PBS was added (30 min, 23 °C). After washing in PBS, the alkaline phosphatase-dependent staining was developed by 0.2 mg naphthol-AS-MX phosphate / ml in combination with 10 mg / ml

TABLE 1 **IDENTIFICATION OF ELASTASE IN SUPERNATANTS OF STIMULATED HUMAN GRANULOCYTES AS THE FUNCTIONAL PROTEASE TO DEGRADE TUMOR CELL pro-uPA**

	FIBRIN CLOT LYSIS (%)	AMIDOLYTIC ACTIVITY (%)
BUFFER	100 +/- 3.8	100 +/- 15
+ CYTOCHALASIN B	104.3 +/- 13.8	91.3 +/- 28.8
+ CYTOCHALASIN B AND FNLPNTL	12.8 +/- 2.6	13.1 +/- 6.9
+ CYTOCHALASIN B AND FNLPNTL, THEN ADDITION OF EGLIN C	102.1 +/- 3.2	131.3 +/- 16.9
+ CYTOCHALASIN B AND FNLPNTL, THEN ADDITION OF ANTI-BODY TO ELASTASE	100 +/- 3.2	98.8 +/- 17.5

5×10^6 granulocytes / ml PBS containing 0.5 mM $MgCl_2$, 1 mM $CaCl_2$, and 0.1% BSA were pretreated with 5 µg cytochalasin B / ml (5 min, 37 °C) prior to the addition of 10^{-8} M FNLPNTL (30 min, 37 °C). Supernatants were collected by low speed centrifugation and then incubated with eglin C (12 µM) or monoclonal antibody to elastase (15 µM) for 30 min, 37 °C. Cells which had been treated with buffer or cytochalasin B, only, served as controls. Aliquots of 30 µl of pro-uPA (500 ng) were mixed with 30 µl of the various cell supernatants (30 min, 37 °C) and then subjected to fibrin clot assay as described in the Materials and Methods section. For assay of amidolytic activity with substrate S-2444, 0.3 nM of plasmin was added. The reaction was stopped by addition of 200 KIU aprotinin / ml. Average values obtained from 4 different donors are expressed +/- standard deviation.

TABLE 2 **Detection of elastase but not uPA in supernatants of granulocytes stimulated with the chemotactic peptide FNLPNTL**

	Elastase		uPA	
	Activity (U/ml)	Antigen (µg/ml)	Activity (U/ml)	Antigen (ng/ml)
Buffer	ND	0.02 +/- 0.02	ND	0.29 +/- 0.11
Cytochalasin B (5 µg / ml)	0.016 +/- 0.01	3.2 +/- 3.3	ND	0.21 +/- 0.03
Cytochalasin + 10^{-8} M FNLPNTL	0.26 +/- 0.01	62.5 +/- 48.7	ND	0.28 +/- 0.07

Granulocytes (5×10^6 / ml) were preincubated with cytochalasin B (5 µg / ml) and then stimulated with 10^{-8} M FNLPNTL (30 min, 37 °C) The cell supernatants were assayed for elastase and uPA activity und antigen. Values (n=4) were expressed +/- standard deviation. <u>Note</u>: Human granulocytes contain uPA **(18)**. Under the conditions applied no significant release of uPA was detected by ELISA or enzymatic assay. **ND:** not detected.

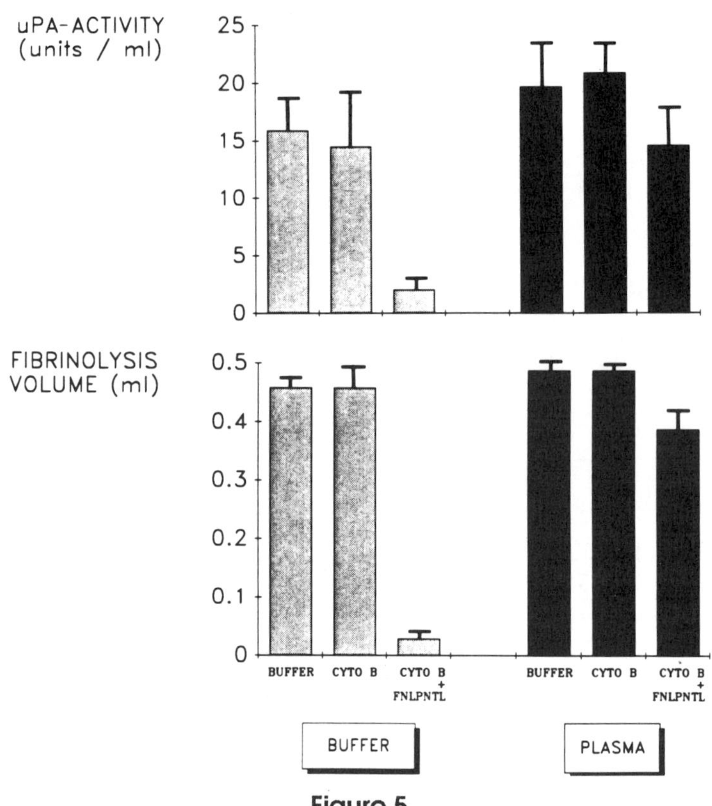

Figure 5

Effect of granulocyte supernatants on pro-uPA

Determination of amidolytic **(upper graph)** and fibrinolytic **(lower graph)** activity. Supernatants (60 µl) obtained from granulocytes treated with the reagents indicated were mixed with 100 units of pro-uPA contained in 60 µl PBS or plasma (30 min, 37 °C). One aliquot (60 µl) was applied to a fibrin clot. The other aliquot (60 µl) was treated with 2.8 nM of plasmin for 45 min at 37 °C and subsequently amidolytic activity was measured with substrate S-2444. Untreated HMW-uPA (prepared from pro-uPA by plasmin treatment) served as control. Black bars show results obtained with pro-uPA in 50 % plasma, grey bars in buffer, only.

Fast Red TR in 0.2 M Tris-HCl, pH 8.5, containing 1 mM levamisole to block intrinsic alkaline phosphatase activity (1 h, 23 °C). The tissue sections were washed in PBS and water and mounted in glycerol-gelatine. Controls were perfomed by using irrelevant mouse-IgG-antibodies and by preabsorbing the antibodies with excess of purified antigen.

Areas selected in **Fig. 6 and 7** represent central part of the tumor. Immunohistochemical staining (dark color, originally red) for the presence of uPA in tumor cells **(Fig. 6)** or elastase in phagocytes **(Fig. 7,** see arrows). The elastase containing cells were located in the tumor stroma and also close to tumor cells. Nuclei were stained with hematoxylin. Microscopic magnification: x 400.

DISCUSSION

Phagocytic cells are important mediators of tissue-destroying events in inflammation but also in tumor spread **(19)**. Tumor cells and phagocytic cells coexist in cancer tissue **(20-24)**. Granulocytes contain elastase, collagenase, and gelatinase. These proteases are released by chemotactic activation of the cells and are linked to the phagocyte's ability to release toxic agents upon receptor-mediated stimulation which will lead to dissolution of the extracellular matrix and destruction of cells **(25)**. Phagocytic cells were located in the tumor tissue and the extracellular matrix (tumor stroma) surrounding the tumor nests and identified by immunohistochemistry **(26)**. One of the functions attributed to granulocytes accumulated in tumor tissue is the cytotoxic effect on the tumor cells by releasing reactive oxidative intermediates upon stimulation. These oxidants are supposed to destroy tumor cells directly **(27,28)**.

Dvorak et al. observed that granulocytes can attach to tumor cells and then elastase-containing granules may be released **(29)**. The concentration of protease inhibitors in such regions should be lower than in plasma and would not be sufficient to neutralize all the elastase activity. Studies by Gudewicz and Gilboa **(30)** demonstrated that enzymatically active urokinase-type plasminogen activator (HMW-uPA) is a potent chemotactic factor for granulocytes <u>in vivo</u> in addition to tumor stroma breakdown products which may also attract granulocytes into the tumor tissue. Cancer cells in solid tumors secrete pro-uPA **(1,2,4)**.

The conversion of enzymatically inactive pro-uPA to active HMW-uPA is limited to the specificity of the protease involved. Plasmin converts pro-uPA into enzymatically active HMW-uPA **(1-3)**. Trypsin and plasma kallikrein may substitute for plasmin **(6)**. Proteolytic action of thrombin **(6,7),** and as shown in the present report, granulocyte elastase, on pro-uPA results in enzymatically inactive uPA-forms which cannot be activated anymore **(17)**. Plasmin and trypsin specifically cleave the Lys^{158}-Ile^{159} peptide bond within pro-uPA **(6)**. Thrombin specifically cleaves at the Arg^{156}-Phe^{157} position which is located two amino acid residues prior to the plasmin cleavage site **(6)**.

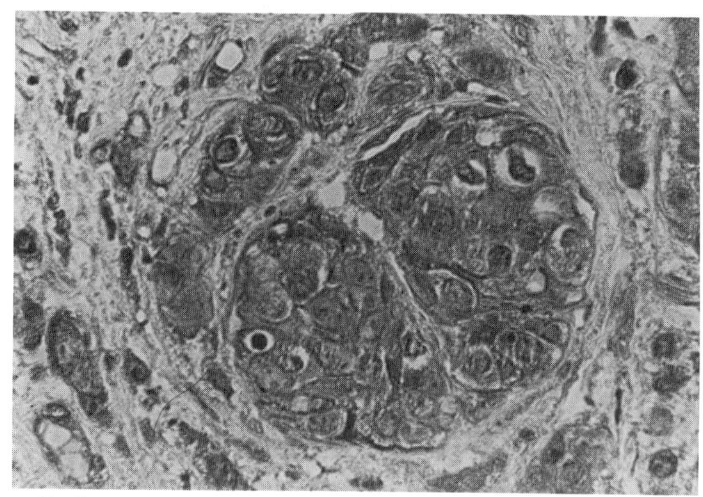

Figure 6

Localization of uPA in tumor cells of breast cancer tissues

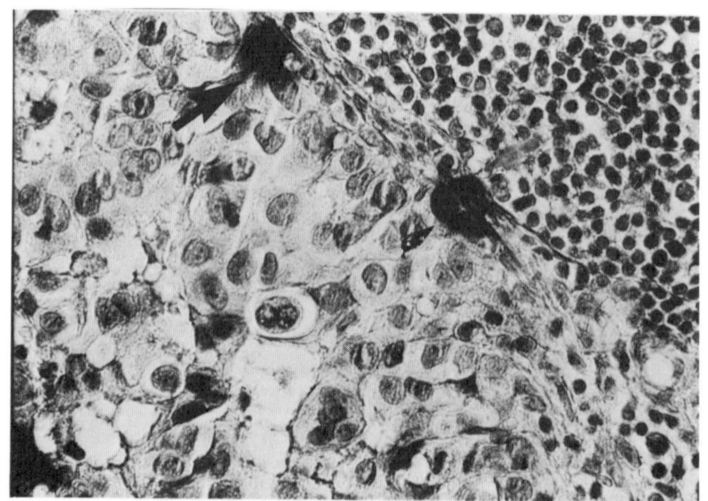

Figure 7

Localization of elastase in phagocytic cells of breast cancer tissues

We presented evidence that human granulocyte elastase specifically cleaves pro-uPA at the peptide bonds Ile[159]-Ile[160] and Thr[165]-Thr[166], the former cleavage site beeing the predominant **(17)**. These cleavage sites are located just behind the plasmin cleavage site and yield A- and B-chains of similar length as shown by SDS-PAGE for plasmin degradation. Cleavage of pro-uPA by proteases will cause a conformational change within the molecule **(6,7,31,32)**. Depending which peptide bond is cleaved, an enzymatically active or inactive two-chain molecule is generated. On the other hand, enzymatically active HMW-uPA (which has been formed by previous plasmin treatment of pro-uPA) is not inactivated by subsequent elastase **(17,31)** or thrombin **(7)** action.

Although granulocyte elastase does not inactivate HMW-uPA, the enzymatic activity of HMW-uPA in solution or bound to tumor cells may be inhibited by the specific plasminogen activator inhibitors PAI-1 and PAI-2 **(1,2,5)**. HMW-uPA bound to receptors on tumor cells is supposed to be involved in tumor growth, malignant invasion, degradation of the tumor stroma and metastasis **(1,2,9,11)**. Inactivation of pro-uPA by granulocyte elastase, in addition to inactivation of HMW-uPA by PAI-1/2, therefore may delay the proteolytic processes in tumor-associated fibrinolysis **(1,9)**. Elastase was identified by us in cell supernatants of chemotactically activated human granulocytes as the functional enzyme to inactivate pro-uPA <u>in vitro</u> as both, eglin C and a monoclonal antibody to elastase, blocked cleavage of pro-uPA by these cell supernatants. **(17)**. Heiple and Ossowski also reported on the destructive potential of granulocyte supernatants on purified pro-uPA. These authors stimulated granulocytes with immune complexes but did not identify the proteases released **(18)**. The inactivation capacity of the cell supernatants was inhibited by the serine protease inhibitor DFP. Among other proteases, elastase is inactivated by DFP. Heiple's and our results support the notion that elastase is one of the key enzymes in inactivating pro-uPA released by phagocytic cells. <u>In vivo</u> inactivation of pro-uPA by elastase may occur in the extravascular space, which is low in elastase inhibitors such as the alpha-1-protease inhibitor.

In this report we demonstrate that breast cancer tissues contain granulocytes which stain for elastase and tumor cells which stain for uPA. In <u>in vitro</u> experiments we have shown that purified tumor cell pro-uPA is degraded by granulocyte elastase. Tumor stroma proteins such as fibrin-(ogen), fibronectin, elastin and collagen are also substrates for granulocyte elastase **(33,34)**. We show that in plasma pro-uPA is only slightly inactivated by elastase indicating inhibition of elastase activity by plasma inhibitors. Weitz et al. **(35)** stated that fibrinopeptide A-alpha$_{1-21}$, which was detected in plasma, is a specific product of elastase action on fibrinogen. This finding suggests that a certain amount of active elastase does exist in the blood circulation. The concentration of uPA/pro-uPA in plasma of healthy donors and also in breast cancer patients is ~1 ng / ml (0.04 nM) **(36,37)**. The concentration of elastase in plasma is ~23 ng / ml (0.73 nM) **(38)**, which is in excess over uPA/pro-uPA.

The finding that granulocyte elastase can degrade tumor pro-uPA has important implications for tumor biology. Inactivation of pro-uPA by elastase prevents generation of enzymatically active HMW-uPA and thus diminishes tumor cell metastasis and invasion. In this context it is worth to mention that the receptor-binding domain within pro-uPA/uPA (amino acid sequence 20-30) is not degraded by elastase. Elastase-treated pro-uPA still binds to uPA-receptors on tumor cells **(Schmitt, Chucholowski, Kobayashi, unpublished results)**. It has been reported by Kirchheimer et al. that proliferation of tumor cells via binding of uPA to surface receptors may occur **(39)**. Stimulation of breast cancer cells with protease-treated pro-uPA will tell whether stimulation of tumor cells is mediated by binding of uPA_{20-30}, only. Internalization of the uPA/uPA-receptor complex (which may be the trigger for cell stimulation) is possible only when enzymatically active HMW-uPA has reacted with the uPA-inhibitor PAI-1 **(5)**.

ABBREVIATIONS

APAAP	alkaline-phosphatase-anti-alkaline-phosphatase complex
BSA	bovine serum albumin
DFP	diisopropylfluorophosphate
FNLPNTL	N-formyl-norleucyl-leucyl-phenylalanyl-norleucyl-tyrosinyl-lysin
HMW-uPA	enzymatically active high-molecular-weight two-chain form of urokinase-type plasminogen activator
moAB	monoclonal antibody
PAP	phosphatase-anti-phosphatase complex
pro-uPA	proenzyme form of the urokinase-type plasminogen activator
S-2444	pyro-Glu-Gly-Arg-p-nitroanilide
SDS-PAGE	sodium dodecyl sulfate polyacrylamide gel electrophoresis
TBS	Tris-buffered saline, pH 8.5
TMB	3,3',5,5'-tetramethylbenzidine.

ACKNOWLEDGEMENT

This work was supported by **Sonderforschungsbereich 207 (Project G10)** of the Deutsche Forschungsgemeinschaft and a grant of the **Wilhelm Sander-Stiftung**. The technical expertise of Mrs. I. Hoepner, B. Jaud-Münch, and E. Sedlaczek is highly acknowledged. The authors thank Dr. W. Gössner, Pathologisches Institut der Technischen Universität München, for providing the fixed breast tumor tissue sections. Dr. Kanayama was a visiting investigator supported by the Japanese Educational Ministry.

REFERENCES

1) Dano, K., Andreasen, P.A., Grondahl-Hansen, J., Kristensen, P., Nielsen, L.S., and Skriver, L.: Plasminogen activators, tissue degradation, and cancer. Adv. Cancer Res. 44, 139 (1985).

2) Blasi, F.: Surface receptors for urokinase plasminogen activator. Fibrinolysis 2, 73 (1988).

3) Günzler, W.A., Steffens, G.J., Öting, F., Kim, S.M.A., Frankus, E., and Flohé, L.: The primary structure of high molecular mass urokinase from human urine. Hoppe-Seyler's Z. Physiol. Chem. 363, 1155 (1982).

4) Markus, G.: The relevance of plasminogen activators to neoplastic growth. A review of recent literature. Enzyme 40, 158 (1988).

5) Cubellis, V., Andreasen, P., Ragno, P., Mayer, M., Dano, K., and Blasi, F.: Accessibility of receptor-bound urokinase to type-1 plasminogen activator inhibitor. Proc. Natl. Acad. Sci. (USA). 86, 4828 (1989)

6) Ichinose, A., Fujikawa, K. and Suyama, T.: The activation of pro-urokinase by plasma kallikrein and its inactivation by thrombin. J. Biol. Chem. 261, 3486 (1986).

7) Gurewich, V. and Pannell, R.: Inactivation of single chain urokinase (pro-urokinase) by thrombin and thrombin-like enzymes: relevance of the findings to the interpretation of fibrin-binding experiments. Blood 69, 769 (1987).

8) Wilhelm, O., Hafter, R., Coppenrath, E., Pflanz, M., Schmitt, M., Babic, R., Linke, R., Gössner, W., and Graeff, H.: Fibrin-fibronectin compounds in ovarian tumor ascites and their possible relation to the tumor stroma. Cancer Res. 48, 3507 (1988).

9) Laemmli, U.K.: Cleavage of structural proteins during the assembly of the head of bacteriophage T4. Nature 277, 680 (1970).

10) Jänicke, F., Schmitt, M., Hafter, R., Hollrieder, A., Babic, R., Ulm, K., Gössner, W., and Graeff, H.: The urokinase-type plasminogen activator (u-PA) is a potent predictor of early relapse in breast cancer. Fibrinolysis 4, 69 (1990).

11) Hollrieder, A., Schmitt, M., Jänicke, F., Kanayama, N., Hafter, R., Gössner, W., and Graeff, H.: Urokinase-type plasminogen activator (uPA) in human breast cancer cells: Immunological localization and quantitation by ELISA applying the same monoclonal antibody. J. Cancer Res. Clin. Onc., in press (1991).

12) Friberger, P.: Synthetic peptide substrate assays in coagulation and fibrinolysis and their application on automates. Semin. Thromb. Hemostas. 9, 281 (1983).

13) Kruithof, E.K.O., Ransijin, A. and Bachmann, F.: Influence of detergents on the measurement of the fibrinolytic activity of plasminogen activators. Thromb. Res. 28, 251 (1982).

14) Boyum, A.: Separation of leucocytes from blood and bone marrow. Scand. J. Clin. Lab. Invest. 21, 77 (1968).

15) Edman, P., and Henschen, A.: Sequence determination. In: Protein Sequence Determination (Needleman, S.B., ed.). 2nd edition, Springer, Berlin, p. 244 (1986).

16) Schmitt, M., Kanayama, N., Henschen, A., Hollrieder, A., Hafter, R., Gulba, D., Jänicke, F., and Graeff, H.: Elastase released from human granulocytes stimulated with N-formyl-chemotactic peptide prevents activation of tumor cell prourokinase. FEBS Lett. 255, 83 (1989).

17) Heiple, J.M. and Ossowski, L.: Human neutrophil plasminogen activator is localized in specific granules and is translocated to the cell surface by exocytosis. J. Exp. Med. 164, 826 (1986).

18) Dvorak, H.F.: Tumors: Wounds that do not heal. Similarities between tumor stroma generation and wound healing. N. Engl. J. Med. 315, 1650 (1986).

19) De Bruin, P.A.F., Griffioen, G., Verspaget, H.W., Verheijen, J.H, Dooijewaard, G., Ingh, H.F., Lamers, C.B.W. Plasminogen activator profiles in neoplastic tissues of the human colon. Cancer Res. 48, 4520 (1988).

20) Markus, G., Takita, H., Camiol, S.M., Corsanti, J.G., Evers, J.L., Hobika, H.: Content and characterization of plasminogen activators in human lung cancer and normal lung tissue. Cancer Res. 40, 841 (1980).

21) Kirchheimer, J.C., Köller, A., Binder, B.R.: Isolation and characterization of plasminogen activators from hyperplastic and malignant prostate tissue. Biochem. Biophys. Acta. 797, 256 (1984).

22) Evers, J.L., Patel, J., Madeja, J.M., Schneider, S.L., Hobika, G.H., Camiolo, S.M., Markus, G.: Plasminogen activator activity and composition in human breast cancer. Cancer Res. 42, 219 (1982).

23) Weiss, J.: Tissue destruction by neutrophils. N. Engl. J. Med. 320, 365 (1989).

24) Larsson, G., Larsson, A. and Astedt, B.: Tissue plasminogen activator and urokinase in normal, dysplastic and cancerous squamous epithelium of the uterine cervix. Thromb. Haemost. 58, 822 (1987).

25) Gudewicz, P.W. and Gilboa, N.: Urokinase (UK) is a chemoattractant for human polymorphonuclear leukocytes. Biochem. Biophys. Res. Comm. 147, 1176 (1987).

26) Sumi, H., Kosugi, T., Toki, N. and Mihara, H.: Elastase-digested urokinase (ED-UK). Experientia 41, 1546 (1985).

27) Bogusky, M.J., Dobson, C.M., and Smith, R.A.G.: Reversible independent unfolding of the domains of urokinase monitored by ^{1}H-NMR. Biochemistry 28, 6728 (1989).

28) Graeff, H., and Hafter, R.: Clinical aspects of fibrinolysis. In: Haemostasis and Thrombosis. A.L. Bloom and D.P. Thomas (eds.). Churchill Livingstone, Edinbourgh, p. 245 (1987).

29) McGowan, S.E., Stone, P.J., Snider, G.L., and Franzblau, C.: Alveolar macrophage modulation of proteolysis by neutrophil elastase in extracellular matrix. Am. Rev. Respir. Dis. 130, 734 (1984).

30) Weitz, J. I., Landman, S. L., Crowley, K. A., Birken, S. and Morgan, F. J.: Development of an assay for in vivo human neutrophil elastase activity. Increased elastase activity in patients with alpha 1-proteinase inhibitor deficiency. J. Clin. Invest.78, 155 (1986).

31) Bachmann, F.: Fibrinolysis. In: Thrombosis and Haemostasis. Verstraete, M., Vermylen, J., Lijnen, H.R., Arnout, J. eds), , Leuven University Press, Leuven, p. 227 (1987).

32) Grondahl-Hansen, J., Agerlin, P., Munkholm-Larsen, P., Bach, F., Nielsen, L.S., Dombernowsky, P., and Dano, K.: Sensitive and specific enzyme-linked immuno-sorbent assay for urokinase-type plasminogen activator and its application to plasma from patients with breast cancer. Lab. Clin. Med. 111, 42 (1988).

33) Plow, E. P.: The contribution of leukocyte proteases to fibrinolysis. Blut 53, 1 (1986).

34) Kirchheimer, J.C., Wojta, J., Christ, G., and Binder, B.R.: Proliferation of a human epidermal tumor cell line stimulated by urokinase. FASEB J. 1:125-1 (1987).

ANCA: A CLASS OF VASCULITIS-ASSOCIATED AUTOANTIBODIES AGAINST MYELOID GRANULE PROTEINS: CLINICAL AND LABORATORY ASPECTS AND POSSIBLE PATHOGENETIC IMPLICATIONS

R Goldschmeding[1], JW Cohen Tervaert[2], KM Dolman[1], AEG Kr von dem Borne[1,3] and CGM Kallenberg[2]

INTRODUCTION

Wegener's granulomatosis (WG) is histologically characterized by extravascular necrotizing granulomatous inflammation, necrotizing and crescentic glomerulonephritis (NCGN) and vasculitis, and primarily affects the upper- and lower airways and the kidney (1, 2). WG is part of the spectrum of systemic necrotizing vasculitides (SNV), which also comprises polyarteritis nodosa (PAN), Churg Strauss syndrome (CSS) and the polyangiitis overlap syndrome (3). The pathogenesis of these diseases is unknown, but immune mechanisms are generally assumed to play a role.

Recently, autoantibodies against neutrophil cytoplasm (ANCA) have been detected in sera from patients with WG and related diseases (4-10). In these original studies, the name "ANCA" (anti-neutrophil cytoplasmic autoantibody) or "ACPA" (anti-cytoplasmic autoantibody) was used exclusively for antibodies producing a characteristic, granular staining of the cytoplasm of ethanol fixed neutrophils and monocytes. Detection of these antibodies was found to have a high specificity and sensitivity for active WG. Moreover, the serum titer of ANCA was found to correlate with disease activity (7-9) and relapses of WG were invariably found to be preceeded by a rise of the ANCA titer (11).

1: Department of Immunohematology, Central Laboratory of the Netherlands Red Cross Blood Transfusion Service, Amsterdam, The Netherlands.
2: Department of Clinical Immunology, State University Hospital Groningen, The Netherlands.
3: Department of Hematology, Academic Medical Center, Amsterdam, The Netherlands.

New Aspects of Human Polymorphonuclear Leukocytes
Edited by W.H. Hörl and P.J. Schollmeyer, Plenum Press, New York, 1991

The importance also of antibodies producing (artifactual) perinuclear staining (fig.1b) was recognized later on (12-14), and the terms "C-ANCA" and "P-ANCA" were introduced (15).

C-ANCA and P-ANCA have both been detected in patients with WG, limited WG, and idiopathic NCGN (4-22). Despite this overlap between C-ANCA and P-ANCA related diseases, there are also clear differences between clinical and histological manifestations found in association with either type of ANCA (17, 20-22).

Understanding of the possible role of ANCA in WG, NCGN and other forms of SNV, will require identification of target antigens and detailed analysis of clinical and histological findings in relation to different antigenic specificities of C- and P-ANCA.

IDENTIFICATION OF C-ANCA ANTIGEN

The originally described WG-specific C-ANCA were found to have an apparently uniform specificity, as judged by Western blots, radio immunoprecipitation, and antigen catching ELISA.

These C-ANCA recognize a N-glycosylated protein of Mr 29K in the primary lysosomes of neutrophils and monocytes, as was established by immunoprecipitation and Western blot analysis of total cell extracts (1% Nonidet P-40), subcellular fractionation (13, 14, 23-25) and by double-labeling immunoelectron microscopy (26). Immuno(histo)logically, the C-ANCA antigen can be detected in human and chimpanzee neutrophils and monocytes, but not in other blood cells or tissues, and not in less related species (23).

In SDS gels, the "29K C-ANCA-antigen" actually migrates as 3 distinct glycoprotein bands of Mrs 29K, 30.5K and 32K. The relative amount of radioactivity in these individual bands was approximately the same when immunoprecipitates obtained with 7 different (WG-) patients' sera and with a monoclonal antibody against the 29K C-ANCA antigen (McAb 12.8) were analyzed semiquantitatively (14). Probably the 3 bands represent isoforms of the same protein, which is also suggested by the finding of a single N-terminal sequence for affinity-purified C-ANCA antigen (see below). There was no change in electrophoretic mobility upon reduction of disulphide bridges. However, antigenicity in Western blots was completely destroyed by such treatment (27, 28, and own unpublished observation), which indicates that conformational determinants are involved in immune recognition of the antigen. Treatment of immunoprecipitated C-ANCA antigen with N-glycanase caused a reduction of its Mr with about 1 kDa. Specific affinity labeling with tritiated diisopropyl-fluorophosphate (^{3}HDFP) showed that the 29K C-ANCA antigen is a serine protease (14, 24). We have purified this serine protease by immunoaffinity chromatography and reversed-phase HPLC. Subsequently, a single N-terminal amino acid sequence was obtained which showed homology to other members of the serine protease family. The N-terminal sequence of our

purified antigen is virtually identical to that of affinity
purified 29K C-ANCA antigen published by Lüdemann et al. (27,
28) and more recently by Niles et al. (25) and to that of a
protein with antimicrobial activity provisionally named "peak
VII", which was isolated from azurophilic granules by rpHPLC
(29). In figure 1 the N-terminus of C-ANCA antigen as
determined by different groups (25, 27, 33) is compared with
that of the other serine proteases of neutrophil azurophilic
granules (29-32). The N-terminus of C-ANCA antigen is most
homologous with that of proteinase 3 (32-34), an enzyme
recently reported to cause emphysema upon intratracheal
instillation in hamsters (35). This confirms previous
preliminary data (14) and is in accordance with the findings
of Lüdemann et al., who in addition to sequence homology also
demonstrated striking similarity of biochemical and functional characteristics of immuno-affinity purified C-ANCA
antigen and proteinase 3 (27, 28). Moreover, we have observed
that in acid gelelectrophoresis, as originally employed for
identification of proteinase 3 (34), C-ANCA antigen co-
migrates with this enzyme. However, definite proof of the
identity of 29kD C-ANCA antigen with proteinase 3 has thusfar
not been obtained.

IDENTIFICATION OF P-ANCA ANTIGENS

In contrast to C-ANCA sera, P-ANCA sera were found to be
heterogeneous. They may recognize 1 or more of several
different antigens of the primary lysosomes, but also of the
specific granules of myeloid leukocytes. Thusfar we have
identified sera containing anti-MPO, anti-MPO + anti-Elastase, anti-MPO + anti-Elastase + anti-Lactoferrin, anti-
MPO + anti-29k, anti-Elastase, and anti-Elastase + anti-29k
(12, 14). The artifactual perinuclear staining obtained with
these antibodies is due to nucleophilicy of the target
antigens and depends on the fixation procedure used (13, 15).
When a formalin-acetone mixture is used instead of ethanol,
cytoplasmic staining is produced also by these P-ANCA sera.
However, P-ANCA staining can also be obtained with some
anti-nuclear antibodies (ANA), and with "granulocyte-specific
ANA" (GS-ANA), the target antigen(s) of which have thusfar
not been identified (36)).

THE CLINICAL SIGNIFICANCE OF ANCA DETECTION

A. Clinical association of C-ANCA

C-ANCA were first described in 1982 by Davies et al. (4)
in the sera of 8 patients with necrotizing and/or crescentic
glomerulonephritis, in all of them associated with constitu-
tional symptoms and arthralgias and/or myalgias. In addition,
respiratory symptoms were found in four of these 8 patients.
On the basis of retrospectively performed antibody studies, 7
of them were suspected of a previous Ross River virus
infection. Two years later, Hall et al. (5) described C-ANCA
in four patients with multisystem disease including lung
involvement in all four patients, necrotizing glomerulonephritis in three, and skin vasculitis in two of these
patients. The condition was labeled 'arteritis'. In 1985,

Rasmussen et al. (6) and van der Woude et al. (7) described
C-ANCA in 25 out of 27 patients with active, biopsy confirmed
Wegener's Granulomatosis. In the latter study, C-ANCA were
not detected in classical polyarteritis, small-vessel
vasculitis, temporal arteritis, other granulomatous disorders
or connective tissue diseases such as systemic lupus erythe-
matosus and rheumatoid arthritis (7). The specificity of C-
ANCA for WG was confirmed in 1986 by Gross et al. (8) who
found C-ANCA in each of their 18 patients with active generalized WG but not in 730 healthy and 265 disease controls.

From these early studies the diagnostic problems of
patients with C-ANCA became apparent: the first reported 12
patients (4, 5) were labeled microscopic polyarteritis or
'arteritis', while WG was diagnosed in the next 43 patients
with positive C-ANCA (6, 7). Nowadays, it is apparent that C-
ANCA can be found in 84%-96% of patients with active, generalized, biopsy proven WG (9, 11, 16, 37, 38) and in 67%-86% of
patients with active locoregional or limited disease (19, 37,
38).

C-ANCA, however, is also detected in the sera of
patients with symptoms highly suggestive of WG, in whom, in
spite of extensive investigation, no WG-specific histological
diagnosis can be made (18, 37-41). In addition, C-ANCA can be
found in 67% of patients with necrotizing and/or crescentic
glomerulonephritis with systemic manifestations but without
granulomatous inflammation of the respiratory tract (9, 13,
20), and in 25% of patients with idiopathic necrotizing
and/or crescentic glomerulonephritis (13, 20). The first
condition is often designated (at least in the United King-
dom) as microscopic polyarteritis (MPA). This term is used to
describe patients with necrotizing and/or crescentic glomerulonephritis without definite clinical or pathological
evidence of WG, but with clinically suspected or biopsy
proven small vessel vasculitis (42).

The occurrence of C-ANCA has also been reported in some
patients with systemic necrotizing vasculitis (SNV) of the
polyarteritis group - i.e. classical polyarteritis nodosa
(PAN) (38), the Churg Strauss Syndrome (CSS) (16), and the
polyangiitis overlap syndrome (21). C-ANCA, however, is only
infrequently found in these latter conditions (21). It may be
argued that the patients with SNV in association with C-ANCA
may have some overlap between polyarteritis and WG (it should
be remembered that WG originally was called 'a borderline
form of polyarteritis nodosa'(43)). Apart from being present
in WG, (idiopathic) CGN and, incidentally, in SNV, C-ANCA has
been found in sporadic cases of lymphomatoid granulomatosis
(11), Takayashu's arteritis (11), Sjögren's disease (9),
anti-glomerular basement membrane nephritis (44), atrial
myxoma (45), viral enteritis (46), and carcinoma of the lung
(46). However, it is amazing that the few available studies
on large groups of patients still report a specificity of C-
ANCA for WG of 97-99% (11, 38).

In addition, the quantitation of C-ANCA might be used
for assessing or predicting disease activity in patients with
WG (11). In a prospective study, we demonstrated that all

relapses were preceded by a significant rise of C-ANCA levels
(11). The specificity of this finding was shown by the
observation that a significant rise of C-ANCA levels was
followed by a relapse in 92% of the patients (11).

B. Clinical associations of P-ANCA

In the ANCA immunofluorescence test staining patterns
different from that produced by C-ANCA have been recognized;
in particular, a perinuclear pattern (P-ANCA) has been
observed. The target antigen, however, is still unknown for a
large proportion of P-ANCA sera (21). The presence of P-ANCA
is not specific for any disease and the antibodies can be
found in many disorders (unpublished observations). In
contrast, ANCA with specificity for myeloperoxidase (MPO-
ANCA), which accounts for about 10-15% of the sera positive
for P-ANCA (21), is associated with particular disease
patterns. In the first reported series (13, 20) MPO-ANCA was
associated with idiopathic and vasculitis-associated necrotizing and/or crescentic glomerulonephritis (CGN). Most of
these patients had kidney limited disease (15, 20). A substantial number of them, however, additionally had pulmonary
infiltrates on their chest radiographs and presented with
hemoptysis and dyspnea. As a result, they were classified as
clinically suspected for WG (20). Granulomatous inflammation,
however, was found only infrequently in these patients. Also
extensive nasal mucosal ulceration was extremely rare,
whereas this is a rather common finding in patients with C-
ANCA. In addition, patients with MPO-ANCA appeared to have
fewer (and less severe) organ involvement than patients with
C-ANCA (22). Thus, while both MPO-ANCA and C-ANCA are associated with idiopathic and/or vasculitis associated CGN, MPO-
ANCA is more associated with the idiopathic variant, while C-
ANCA is more associated with CGN as part of classic WG.
Apart from being detected in the sera of patients with CGN,
we recently reported the presence of MPO-ANCA in 50% of
patients with systemic necrotizing vasculitis of the polyarteritis group, with or without signs of glomerulonephritis
(21). Although a few of these patients lacked pulmonary
involvement, most of them had asthma and/or pulmonary
infiltrates. Some fullfilled the diagnostic criteria for the
Churg Strauss Syndrome, but most patients were classified as
the polyangiitis overlap syndrome (47). In summary, MPO-ANCA
are not only found in idiopathic and/or vasculitis-associated
CGN but also in SNV of the polyarteritis group.

The specificity of MPO-ANCA for SNV and/or idiopathic
CGN was tested in selected groups of patients with well
established forms of vasculitis, glomerulonephritis, granulomatous disorders, and connective tissue diseases. It proved
to be very high: 99% (21). The clinical associations of other
P-ANCAs with known antigenic specificities, e.g. anti-human
leucocyte elastase (HLE-ANCA) and anti-lactoferrin (LF-
ANCA), are not yet known but these autoantibodies seem to
occur very infrequently.

In summary, among the various ANCAs described until now,
only C-ANCA (or 29 kD-ANCA) and MPO-ANCA have well established clinical associations. C-ANCA appears to be as-

sociated with active (limited or generalized) WG while MPO-
ANCA is associated with the whole spectrum of systemic
necrotizing vasculitides (in which idiopathic CGN must be
included). Both ANCAs, however, are very specific for this
group of disorders.

POSSIBLE PATHOGENETIC SIGNIFICANCE OF (C- AND P-) ANCA

In patients with WG, relapses of disease have been noted
often to occur at the time of infection with common pathogens
(48) and antibiotic therapy with trimethoprim-sulfamethoxa-
zole has been reported to be of value in treatment and
prevention of relapses (49). With respect to a possible
pathogenic role of ANCA, this is of interest especially since
the target antigens of ANCA are important mediators in
inflammatory processes.

Relapses of WG are sensitively (100%) and specifically
(92%) predicted by a rise of the ANCA titer (11). Upon
treatment with high dose steroids and cytostatic drugs the
ANCA titer falls again, usually after clinical improvement
has occurred (11, 21). In some cases, high titers of ANCA
were noted to persist for more than a year without any sign
of (recurrence of) disease activity. These observations
suggest that a high titer of circulating ANCA is a necessary,
but in itself insufficient condition for disease activity in
WG. This is compatible with the observation that in "resting"
neutrophils the ANCA antigens are present only intracellularly, and therefore are not accessible to circulating autoantibodies. However, upon stimulation, e.g. when (bacterial)
infection occurs, neutrophils release these antigens and
immune complexation with ANCA can occur (14). Subsequently,
ANCA could cause derailment of the inflammatory process by a
number of different mechanisms. Released antigens might form
immunecomplexes with circulating ANCA. Formation of (soluble)
immunecomplexes would be expected to sustain or even amplify
inflammatory activity. In addition, ANCA might interfere with
physiological inactivation of the target enzymes, which
could lead to excessive tissue damage. Alternatively, binding
of ANCA might inhibit the (enzymatic) function of released
antigens, thus frustrating elimination of the inflammatory
stimulus. Because of their high (positive) charge, released
antigens might also attach to plasma membrane components or
(extra-)cellular structures which then become targets for
circulating ANCA. Binding of ANCA might thus cause damage
to- or induce functional changes in "coated" cells and
tissues. In addition, a local proliferative response of
autoreactive T-cells with specificity for the "ANCA-antigens"
might be involved in the formation of granulomas. Perfusion
studies (50, 51) in rat kidneys have demonstrated localization of MPO and of elastase to the glomerular basement
membrane. If this occurs also in patients with ANCA, then
humoral- as well as (T-)cellular autoimmunity to these
antigens could play a role in the developement of the
typically ANCA-associated "pauci-immune" form of necrotizing
and crescentic glomerulonephritis (NCGN) (13, 17, 20, 52, 53,
54).

The eventual outcome of each of these hypothetical
processes would largely depend on the functional and physicochemical characteristics of the antigen(s) involved, and
different types of ANCA would consequently be expected to
have different disease associations. It has become clear
that C-ANCA (29kD-ANCA) and MPO-ANCA indeed have different
clinical and histopathological disease associations, although
there is a certain overlap (17, 20-22).

In short, we suggest that in the presence of ANCA,
inflammatory stimulation by (common) infectious agents can
lead to excessive damage, either through dysregulation of the
inflammatory response or through (in situ) immune complex
formation. In addition, T-cell reactivity to the autoantigens
recognized by ANCA might be involved in the development of
the typical lesions of ANCA-associated disease as well.

```
----------------------------------------------------------------
                       1       5         10        15        20
C-ANCA antigen         I V G G H E A Q P H S R P Y M A S L Q M R
(CLB)
C-ANCA antigen
(28)                   - - - - - - - - - - - I - -*- - -
p29 (25)               - - - - - - - - - - - X - - - - - - - - -
"peak VII" (33)        - - - - - - - - - - - - - - - - - - - E -
proteinase 3 (29)      - - - - - - - E - R W X - G

azurocidin (33)        - - - - R K - R - R Q F - F L - - I - N
leucocyte elastase     - - - - R R - R - - A W - F - V - - - L -
(34)
cathepsin G (35)       - I - - R - S R - - - - - - - - - Y - - I Q
----------------------------------------------------------------
```

Figure 1. N-terminal sequence of the serine proteases of
neutrophil azurophilic granules

X   represents unassigned residue
*: between the P at position 13 and the Y at position 14 in
    this figure, Lüdemann reports an I at position 14 (ref.28).

CONCLUSIONS

C-ANCA and P-ANCA together form a class of autoantibodies against myeloid-specific granule proteins. C-ANCA
sera have an apparently uniform specificity for the 29kD
serine protease ("proteinase-3"), while P-ANCA sera are
heterogeneous and may recognize 1 or more of at least 4
different antigens, including MPO, neutrophil elastase,
lactoferrin, and occasionally also (in addition) the 29kD C-
ANCA antigen.

Despite a certain overlap, C-ANCA (anti-29kD) and MPO-
ANCA (in our experience the most prominent "real" P-ANCA)
show different clinical and histological associations. In
particular formation of necrotizing granulomas, extensive
extra-renal disease, and organ failure were far more prominent in patients with anti-29kD than in patients with anti-
MPO.

It is tempting to speculate that autoimmunity to myeloid
granule proteins, which is reflected in circulating ANCA,
might have pathogenetic significance by causing derailment of
inflammatory processes in response to certain stimuli, which
might include infection with common pathogens. This view is
supported not only by the consistent observation of increase
of ANCA titers preceeding clinical relapses, but also by
reports of successful prevention of relapses with antimicrobial drugs.

The monoclonal antibodies and purified antigens that are
now becoming available will help to increase understanding of
these phenomena by immunohistological analysis of biopsy
specimens from tissues affected by ANCA-related disease,
investigation of the functional consequences of immune-
complexation of (released) target enzymes, and analysis of T-
cell autoreactivity to the ANCA-antigens.

REFERENCES

1. McCluskey, R.T., Fienberg, R.: Vasculitis in primary vasculitides, granulomatoses and connective tissue diseases.
   Human Pathol. 14: 305-315 (1983).
2. Fauci, A.S., Wolff, S.M.: Wegener's granulomatosis:
   studies in eighteen patients and a review of the literature. Medicine (Baltimore) 52:535-561 (1973).
3. Conn, D.L.: Update on systemic necrotizing vasculitis.
   Mayo Clin. Proc. 64: 535-543 (1989).
4. Davies, D.J., Moran, J.E., Niall, J.F., Ryan, G.B.:
   Segmental necrotizing glomerulonephritis with anti-
   neutrophil antibody: possible arbovirus aetiology?
   Br. Med. J. 285: 206 (1982).
5. Hall, J.B., Wadham, B.M., Wood, C.J., Ashton, V., Adam,
   W.R.: Vasculitis and glomerulonephritis: a subgroup with
   an anti-neutrophil cytoplasmic antibody. Aust. NZ. J.
   Med. 14: 277-278 (1984).
6. Rasmussen, N., Wiik, A.: Autoimmunity in Wegener's granulomatosis. In: J.E. Veldman, B.F. McCabe, E.H. Huizing
   and N. Mygind, eds. Immunobiology, Autoimmunity, Transplantation in Otorhinolaryngology. Kugler, Amsterdam. pp
   207-212 (1985).
7. Van der Woude, F.J., Rasmussen, N., Lobatto, S., Wiik,
   A., Permin, H., Van Es, L.A., Van der Giessen, M., Van
   der Hem, G.K., The, T.H.: Autoantibodies against neutrophils and monocytes: tool for diagnosis and marker for
   disease activity in Wegener's granulomatosis. Lancet
   i:425-429 (1985).
8. Gross, W.L., Lüdemann, G., Kiefer, G., Lehmann, H.: Anti-
   cytoplasmic antibodies in Wegener's Granulomatosis
   (letter). Lancet i:806 (1986).
9. Savage, C.O.S., Jones, S., Winearls, C.G., Marshall,
   P.D., Lockwood, C.M.: Prospective study of radioimmunoassay for antibodies against neutrophil cytoplasm in diagnosis of systemic vasculitis. Lancet i: 1389-1393
   (1987).
10. Lüdemann, G., Gross, W.L.: Autoantibodies against cyto-
    plasmic structures of neutrophil granulocytes in
    Wegener's Granulomatosis. Clin. Exp. Immunol. 69: 350-
    357 (1987).
11. Cohen, Tervaert, J.W., Van der Woude, F.J., Fauci, A.S.,

Ambrus, J.L., Velosa, J., Keane, W.F., Meijer, S., Van der Giessen, M., The, T.H., Van der Hem, G.K., Kallenberg, C.G.M.: Association between active Wegener's Granulomatosis and anticytoplasmic antibodies. <u>Arch. Intern. Med.</u> 149: 2461-2465 (1989).

12. Goldschmeding, R., Cohen, J.W., Tervaert, C.E., Van der Schoot, M., Van der Giessen, A.E.G., Von dem Borne, K.R., Kallenberg, C.G.M.: Autoantibodies against myeloid lysosomal enzymes: a novel class of autoantibodies associated with vasculitic syndromes. <u>Kidney Int.</u> 34:558-559 (1988) (abstract).

13. Falk, R.J., Jennette, J.C.: Anti-neutrophil cytoplasmic autoantibodies with specificity for myeloperoxidase in patients with systemic vasculitis and crescentic glomerulonephritis. <u>New Engl. J. Med.</u> 318:1651-1657 (1988).

14. Goldschmeding, R., Van der Schoot, C.E., Ten Bokkel Huinink, D., Hack, C.E., Van den Ende, M.E., Kallenberg, CGM, von dem Borne AEG Kr: Wegener's granulomatosis autoantibodies identify a novel DFP-binding protein in the lysosomes of normal human neutrophils. <u>J. Clin. Invest.</u> 84:1577-1587 (1989).

15. Jennette, J.C., Falk, R.J.: Anti-neutrophil cytoplasmic autoantibodies. <u>N. Engl. J. Med.</u> 319: 1417 (1988).

16. Harrison, D.J., Simpson, R., Kharbanda, R., Abernathy, V.E., Nimmo, G.: Antibodies to neutrophil cytoplasmic antigens in Wegener's Granulomatosis and other conditions. <u>Thorax</u> 44: 373-377 (1989).

17. Jennette, J.C., Wilkman, A.S., Falk, R.J.: Anti-Neutrophil cytoplasmic autoantibody-associated glomerulonephritis and vasculitis. <u>Am. J. Pathol.</u> 135: 921-930 (1989).

18. Gans, R.O.B., Goldschmeding, R., Donker, A.J.M., Hoorntje, S.J., Kuizenga MC, Cohen Tervaert JW, Kallenberg, C.G.M., Borne AEGKr von dem: Neutrophil cytoplasmic autoantibodies and Wegener's Granulomatosis. <u>Lancet</u> i: 269-270 (1989).

19. Cohen Tervaert, J.W., Goldschmeding, R., Hene, R.J., Kallenberg, C.G.M.: Anti-neutrophil cytoplasmic autoantibodies and Wegener's granulomatosis. <u>Lancet</u> i: 270 (1989).

20. Cohen Tervaert, J.W., Goldschmeding, R., Elema, J.D., Giessen, M. van der, Huitema, M.G., Hem, G.K. van der, The, T.H., Borne, AEGKr von dem, Kallenberg, C.G.M.: Autoantibodies against myeloid lysosomal enzymes in crescentic glomerulonephritis. <u>Kidney Int.</u> 37: 799-806 (1990).

21. Cohen Tervaert, J.W., Goldschmeding, R., Elema, J.D., Limburg, P.C., van der Giessen, M., Huitema, M.G., Koolen, M.I., Hené, R.J., The, T.H., van der Hem, G.K., von dem Borne, AEGKr, Kallenberg, C.G.M.: Autoantibodies to myeloperoxidase are associated with different forms of vasculitis. Submitted.

22. Goldschmeding, R., Cohen Tervaert, J.W., Gans, R.O.B., Dolman, K.M., van den Ende, M.E., Kuizenga, M.C., Kallenberg, C.G.M., von dem Borne, AEG Kr: Different immunological specificities and disease associations of c-ANCA and p-ANCA. <u>Neth. J. Med.</u> 36: 114-116 (1990).

23. Goldschmeding, R., Ten Bokkel Huinink, D., A.E.G. Kr. von dem Borne, T.H. The: Wegener's granulomatosis autoantibodies serological and immunochemical characterization of the antigen. <u>Br. J. Haematol.</u> 66:411 (abstract) (1987).

24. Goldschmeding, R., Ten Bokkel Huinink t, Faber, N., Tetteroo, P.A.T., Vroom, T.M., Hack, C.E., von dem Borne, AEG Kr: Identification of the "ANCA-antigen" as a novel myeloid serine protease. In: Proceedings of the 1st international workshop on ANCA, Copenhagen, january 1988. Rasmussen N, and Wiik A, eds. <u>APMIS</u> 97 suppl. 6:46 (1989).

25. Niles, J.L., McCluskey, R.T., Ahmad, M.F., Arnaout, M.A.: Wegener's Granulomatosis autoantigen is a novel serine proteinase. <u>Blood</u> 74: 1888-1893 (1989).

26. Calafat, J., Goldschmeding, R., Ringeling, P.L., Janssen, H., Schoot, C.E. van der: In situ localisation by double-labeling immunoelectron microscopy of anti-neutrophil cytoplasmic autoantibodies in neutrophils and monocytes. <u>Blood</u> 75: 242-250 (1990).

27. Lüdemann, J., Utecht, B., Gross, W.L.: The target antigen of ACPA is proteinase 3. <u>Neth. J. Med.</u> (Suppl). In Press, (1989).

28. Lüdemann, J., Utecht, B., Gross, W.L.: Anti-neutrophil cytoplasm antibodies in Wegener's Granulomatosis recognize an elastinolytic enzyme. <u>J. Exp. Med.</u> 171: 357-362 (1990).

29. Gabay, J.E., Scott, R.W., Campanelli, D., Griffith, J., Wilde, C., Marra, M.N., Seeger, M., Nathan, C.F.: Antibiotic proteins of human polymorphonuclear leucocytes. <u>Proc. Natl. Acad. Sci. USA</u> 86:5610-5614 (1989).

30. Sinha, S., Watorek, W., Karr, S., Giles, J., Bode, W., Travis, J.: Primary structure of human neutrophil elastase. <u>Proc. Natl. Acad. Sci. USA</u> 84:2228-2232 (1987).

31. Salvesen, G.S., Farley, D., Shuman, J., Przybyla, A., Reilly, C., Travis, J.: Molecular cloning of human cathepsin G: Structural similarity to mast cell and cytolytic lymphocyte proteinases. <u>Biochemistry</u> 26:2289-2293 (1987).

32. Wehner, N.G.: Biochemistry and function of proteases isolated from the cytoplasmic granules of polymorphonuclear leucocytes. Ph.D. Thesis. University of Minnesota, 1987.

33. Dolman, K.M. et al.: (in preparation).

34. Dewald, B., Rindler-Ludwig, R., Bretz, U., Baggiolini, M.: Subcellular localization of neutral proteases in neutrophilic polymorphonuclear leucocytes. <u>J. Exp. Med.</u> 141: 709-723 (1975).

35. Kao, R.C., Wehner, M.G., Skubitz, K.M., Gray, B.H., Hordal, J.R.: Proteinase 3. A distinct human polymorphonuclear leucocyte proteinase that produces emphysema in hamsters. <u>J. Clin. Invest.</u> 82:1963-1973 (1988).

36. Wiik, A.: Granulocyte-specific antinuclear antibodies. <u>Allergy</u> (Copenh.). 35:263-289 (1980).

37. Specks, V., Wheatley, C.L., McDonald, TJMc, Rohrbach, M.S., DeRemee, R.A.: Anticytoplasmic autoantibodies in the diagnosis and follow-up of Wegener's Granulomatosis. <u>Mayo Clin. Proc.</u> 64: 28-36 (1989).

38. Nölle, B., Specks, V., Lüdemann, G.: Anticytoplasmic autoantibodies: their immunodiagnostic value in Wegener's Granulomatosis. <u>Ann. Intern. Med.</u> 111: 28-40 (1989).

39. Parlevliet, K.J., Henzen-Logmans, S.C., Oe, P.L., Bronsveld, W., Balm, A.J.M., Donker, A.J.M: Antibodies to components of neutrophil cytoplasm: a new diagnostic tool in patients with Wegener's Granulomatosis an systemic vasculitis. <u>Q. J. Med.</u> 249: 55-63 (1988).

40. Hoare, T.J., Rhys Evans, P.H.: Anti-neutrophil cytoplasmic antibody assay in diagnosis of recurrent subglottic stenosis. (letter) <u>Lancet</u> ii: 1360 (1988).

41. Cohen Tervaert, J.W., Huitema, M.G., van der Giessen, M., Goldschmeding, R., van der Woude, F.J., Kallenberg, C.G.M.: Wegener's Granulomatosis and anti-cytoplasmic antibodies: the Groningen experience. <u>Acta Pathol. Microbiol. Immunol. Scand.</u> S6, 97: 36 (1989).

42. Savage, C.O.S., Winearls, C.G., Evans, D.J., Rees, A.J., Lockwood, C.M.: Microscopic polyarteritis: presentation, pathology and prognosis. <u>Q. J. Med.</u> 56:467-483 (1985).

43. Klinger, H.: Grenzformen der Periarteritis Nodosa. <u>Frankf. Z. Pathol.</u> 42: 455-480 (1931).

44. Lockwood, C.M., Jayne, D.R., Marshall, P., Jones, S., Savage, C.O.S.: A prospective study of the incidence of anti-GBM and anti-neutrophil cytoplasm antibodies in patients with rapidl progressive nephritis. (abstract). <u>Kidney Int.</u> 33: 329 (1988).

45. Savige, J.A., Yeung, S.P., Davies, D.J., Ebeling, P., Hunt, D.H.: Anti-neutrophil cytoplasmic antibodies associated with atrial myxoma. (letter). <u>Am. J. Med.</u> 85: 755-756 (1988).

46. Venning, M.C., Arfeen, S., Bird, A.G.: Antibodies to neutrophil cytoplasmic antigen in systemic vasculitis. (letter). <u>Lancet</u> ii: 850 (1987).

47. Leavitt, R.Y., Fauci, A.S.: Polyangiitis Overlap Syndrome. <u>Am. J. Med.</u> 81: 79-85 (1986).

48. Pinching, C.A, Rees, A.J., Pussell, B.A., Lockwood, C.M., Mitchison, R.S., Peters, D.K.: Relapses of Wegener's granulomatosis: the role of infection. <u>Br. Med. J.</u> 281: 836-838 (1980).

49. DeRemee, R.A.: The treatment of Wegener's granulomatosis with trimethoprim/sulfamethoxazole: illusion or vision. <u>Arthritis Rheum.</u> 32: 1068-1072 (1988).

50. Johnson, R.J., Couser, W.G., Chi, E.Y., Adler, S., Kleabnoff, S.J.: New mechanisms for glomerular injury. Myeloperoxidase- hydrogen peroxide-halide system. <u>J. Clin. Invest.</u> 79: 1379-1387 (1987).

51. Johnson, R.J., Couser, W.G., Alpers, C.E., Vissers, M., Schulze, M., Klebanoff, S.J.: The human neutrophil serine-proteases, elastase and cathepsin G, can mediate glomerular injury in vivo. <u>J. Exp. Med.</u> 168: 1169-1174.

52. van der Woude, F.J., Daha, M.R., van Es, L.A.: The current status of neutrophil cytoplasmic antibodies. <u>Clin. Exp. Immunol.</u> 78: 143-148 (1989).

53. Rennke, H.G., Klein, Ph.S., Mendrick, D.L.: (abstract). <u>Kidney Int.</u> 37: 428 (1990).

54. Kallenberg, C.G.M., Goldschmeding, R., Cohen Tervaert, J.W., von dem Borne, A.E.G.Kr.: Autoantibodies to myeloid lysosomal enzymes: new clues to vasculitis and glomerulonephritis. A hypothesis based on humoral immune mechanisms. <u>Neth. J. Med.</u> 36: 163-168 (1990).

ANTI-CYTOPLASMIC ANTIBODIES IN WEGENER'S GRANULOMATOSIS ARE

DIRECTED AGAINST PROTEINASE 3*

Jens Lüdemann, Bert Utecht and Wolfgang L. Gross

Abteilung für klinische Rheumatologie der
Medizinischen Universität zu Lübeck und
Rheumaklinik Bad Bramstedt
Federal Republik of Germany

INTRODUCTION

Wegener's granulomatosis (WG), a systemic necrotizing granulomatous vasculitis, used to be considered a relatively rare
disease, but in the past few years it has been diagnosed much
more frequently. This is at least partially due to the study by
Van der Woude et al. (1985), who showed that autoantibodies directed against a cytoplasmic antigen of human neutrophil granulocytes and monocytes (anti-cytoplasmic antibodies = ACPA, syno-
nym: ANCA; C-ANCA) are a disease-specific marker of WG. The immunodiagnostic value of ACPA has been confirmed by many investigators (Gross et al., 1986; Savage et al., 1987; Lüdemann and
Gross, 1987; Parlevliet et al., 1988). Moreover, it has been
shown that changes in ACPA titer parallel changes in disease
activity (Van der Woude et al., 1985; Parlevliet et al., 1988;
Lüdemann et al., 1988a; Specks et al., 1989; Nölle et al.,
1989).

The correlation between ACPA titer and disease activity
suggests that the role of ACPA is more than just that of an
epiphenomenon (Gross, 1989). Therefore, identification of the
target antigen of ACPA is of crucial importance for further
investigations into the pathogenesis of WG. Initially it was
reported that alkaline phosphatase is associated with the ACPA
antigen (Lockwood et al., 1987). But it had already been shown
earlier (Gross et al., 1987; Goldschmeding et al., 1987; Rasmussen et al., 1987; Lüdemann et al., 1988b) that the target antigen of ACPA is not alkaline phosphatase. Later, myeloperoxidase
was described as an antigen recognized by autoantibodies that
produced artifactual perinuclear immunostaining of ethanol-fixed
neutrophils (Falk and Jennette, 1988). Jennette and Falk (1988)
proposed calling these autoantibodies perinuclear-pattern ANCA

---

* This work was supported by the "Bundesministerium für For-
schung und Technologie" (grant 01 VM 8622) and the "Verein zur
Förderung der Erforschung und Bekämpfung rheumatischer Erkran-
kungen Bad Bramstedt e.V."

New Aspects of Human Polymorphonuclear Leukocytes
Edited by W.H. Hörl and P.J. Schollmeyer, Plenum Press, New York, 1991

(P-ANCA), to distinguish them from the originally described autoantibodies (Van der Woude et al., 1985; Rasmussen et al., 1988), which produce diffuse cytoplasmic immunostaining (ACPA; C-ANCA). Not only myeloperoxidase, but also elastase (Goldschmeding et al. 1989a) and lactoferrin (Thompson and Lee, 1989) have been shown to be target antigens of autoantibodies that induce P-ANCA staining of neutrophils. Recently, Goldschmeding et al. (1989b) reported that ACPA specifically associated with WG (C-ANCA) are directed against a novel lysosomal serine proteinase. We have confirmed these findings. Moreover, we were able to show that ACPA producing the typical cytoplasmic immunostaining (C-ANCA) recognize a conformational epitope of proteinase 3, an elastinolytic neutral serine proteinase (Lüdemann et al., 1990). Here we report our results concerning the characterization of the target antigen of ACPA.

BIOCHEMICAL CHARACTERIZATION OF THE ACPA ANTIGEN

The target antigen of ACPA was released from human neutrophils during degranulation induced by phorbol myristate acetate (Gross et al., 1987) and was further purified by affinity chromatography using a column with bound IgG from an ACPA-positive serum, as described previously (Lüdemann et al., 1988b). The affinity-purified antigen was subjected to different pH values and temperatures and after neutralization, or cooling, it was coated to a microtiter plate. Using the previously described ELISA technique (Lüdemann et al., 1988b) we were able to demonstrate that the antigenicity is destroyed below pH values of approximately 3 and at temperatures of higher than approximately 56°C but not up to pH values of 12. Therefore, separation techniques using buffers with pH values below 3 cannot be employed for antigen preparation.

Under reducing conditions SDS-PAGE revealed that the affinity-purified antigen comprises 3 isoforms with molecular weights beween 26 kD and 28 kD. In immunoblotting, ACPA-positive sera showed no reaction with any of the reduced protein bands, but reacted with a relatively diffuse band in the molecular weight range of 35 kD to 39 kD when the antigen was separated under nonreducing conditions (Lüdemann et al., 1990). The observation that the immunoreactivity of the antigen was totally abolished when disulfide bonds were cleaved by reduction with 2-mercaptoethanol, which even occurred without heating, indicates that the autoantibodies are directed against conformational epitopes of the protein.

The N-terminal amino acid sequence was determined from antigen that was separated in SDS-PAGE under nonreducing conditions and transferred electrophoretically to a polyvinylidene difluoride membrane using the method described by Matsudaira (1987). The first sequence analysis revealed 9 amino acids and the second determination with antigen prepared from different donors extended the sequence information to the 17 N-terminal amino acids Ile-Val-Gly-Gly-His-Glu-Ala-Gln-Pro-His-Ile-Arg-Pro-Lle-Tyr-Met-Ala (Lüdemann et al., 1990). This sequence was not identical with any of the sequences listed in the January 1989 release of the MIPSX database (Martinsrieder Institut für Proteinsequenzen, Martinsried, FRG), which contains all available protein and nucleic acid databases. However, some considerable homologies with known serine proteinases were detected, especially the highly conserved N-terminal sequence Ile/Val-Ile/Val-Gly-Gly in connection with Pro in position 13. Interestingly, the two well characterized serine proteinases of human neutrophils, elastase (Shina et al., 1987) and cathepsin G (Heck et al., 1986) as well as the recently described proteinase 3 (Wehner, 1987) share considerable N-terminal sequence homology with the antigen.

FUNCTIONAL CHARACTERIZATION OF THE ACPA ANTIGEN

Goldschmeding et al. (1989b) were able to immunoprecipitate a 29 kD protein labeled with tritiated diisopropyl fluorophosphate (inhibitor of serine proteinases) using ACPA-positive sera. Their results, together with the sequence data, encouraged us to investigate the enzymatic activity of the antigen. Initially we found that the antigen is able to cleave alpha-naphthyl acetate, a substrate that is hydrolysed by many proteinases. The pH optimum for esterase activity with α-naphthyl acetate was between 6.75 and 7.25 (Lüdemann et al., 1990), indicating that the antigen is a neutral proteinase. The specific substrates for human leukocyte elastase (Suc-Ala-Ala-Ala-pNA, Suc-Ala-Ala-Val-pNA and MeO-Suc-Ala-Ala-Pro-Val-pNA), for cathepsin G (Suc-Ala-Ala-Pro-Phe-pNA) and for trypsin (Tosyl-Arg-Methyl-ester) were not hydrolysed by the antigen, but hemoglobin was degraded to a considerable extent (Table 1). Moreover, we showed that the antigen is able to cleave elastin (Lüdemann et al., 1990).

The conclusion drawn from the sequence data, that the target antigen of ACPA is a serine proteinase, was further substantiated by utilizing inhibitors with relative enzyme class specificity. The α-naphthyl acetate esterase activity of the antigen was not inhibited by pepstatin (inhibitor of carbonyl proteinases), N-ethylmaleimide (inhibitor of thiol proteinases) or EDTA (inhibitor of metalloproteinases), but by phenylmethylsulfonyl fluoride (PMSF), the specific inhibitor of serine proteinases (Table 2). In addition, we investigated whether the elastinolytic activity of the antigen is inhibited by alpha-2-macroglobulin (α-2-M) or alpha-1-proteinase inhibitor (α-1-PI; formerly: alpha-1-antitrypsin). α-2-M is a relatively nonspecific inhibitor of many different proteinases. α-1-PI is specific for serine proteinases and is suspected to be the most important biological

Table 1. Substrate Specificity of the ACPA Antigen Compared with Human Neutrophil Elastase (HLE) and Cathepsin G

Substrate	ACPA-Antigen	HLE	Cathepsin G
α-naphthyl acetate	100	100	100
Suc-Ala-Ala-Ala-pNA	0	100	0
Suc-Ala-Ala-Val-pNA	0	100	0
MeO-Suc-Ala-Ala-Pro-Val-pNA	0	100	0
Suc-Ala-Ala-Pro-Phe-pNA	0	0	100
Tosyl-Arg-Methyl-ester	0	ND	ND
Hemoglobin	97	60	100

Degradation of the different substrates was assayed as described by Kao et al. (1988). The values for substrate specificity are expressed as percentage of the enzyme most active with each substrate (100%). ND = not determined.

Table 2. Effects of Potential Inhibitors on Alpha-Naphthyl
          Acetate Esterolytic Activity of the ACPA Antigen
          Compared with Human Neutrophil Elastase (HLE) and
          Cathepsin G

Inhibitor	Final Concentration	Esterolytic Activity (%)		
		ACPA Antigen	HLE	Cathepsin G
None	−	100	100	100
Pepstatin	0.01 mM	100	100	92
N-ethylmaleimide	1 mM	99	98	97
EDTA	5 mM	94	102	86
PMSF	1 mM	4	0	0

Enzymatic activity is expressed as percentage of the proteinase-
induced naphthol release in the absence of inhibitors. Inhibitors were preincubated with the enzymes for 30 min at room temperature before determining the α-naphthyl acetate esterolytic
activity (according to Kao et al., 1988).

inhibitor of human leukocyte elastase. The elastinolytic activity of the antigen was inhibited by both α-2-M and α-1-PI, indicating that these important human proteinase inhibitors are
also physiological inhibitors of proteinase 3.

COMPARISON OF THE ACPA ANTIGEN AND PROTEINASE 3

     The main features of the ACPA antigen we investigated and
the features of proteinase 3 (Kao et al., 1988) are summarized
in Table 3. Since the molecular weight of the three isoforms of
the antigen, its substrate specificity, its pH optimum with α-
naphthyl acetate and its inhibitor profile are identical with
those reported for proteinase 3 (Kao et al., 1988), we conclude
that ACPA are directed against proteinase 3, the third neutral
serine proteinase of human neutrophils (Baggiolini et al.,
1978). This conclusion was further substantiated by purifying
the antigen with the same chromatographic techniques described
by Kao et al. (1988) for isolation of proteinase 3. The differences in the N-terminal amino acid sequence (Wehner, 1987; Lüdemann et al., 1990) could be due either to microheterogeneity or
more probably to inaccuracies in the sequence analyses.

     To investigate if proteinase 3 is the only target antigen
of ACPA inducing cytoplasmic staining of neutrophils in indirect
immunofluorescence (C-ANCA), we analyzed C-ANCA-positive sera
of 296 patients in our ELISA (Lüdemann et al., 1988b) with affinity-purified antigen (proteinase 3) on the solid phase. Sera
of only 14 patients with C-ANCA did not react with proteinase 3
in the ELISA, indicating that the autoantibodies of at least 95%
of our C-ANCA-positive patients are directed against proteinase 3. Although it cannot be excluded that individual patients
have autoantibodies to more than one antigen, the good correlation between immunofluorescence titers and the ACPA values measured in our ELISA (Lüdemann et al., 1988b) strongly suggests
that in these patients only autoantibodies against proteinase 3
are involved. But on the other hand, it is quite obvious that
a small percentage of C-ANCA recognize one or more neutrophil
cytoplasmic antigens different from proteinase 3.

Table 3.   Comparison of the ACPA Antigen and Proteinase 3

	ACPA antigen	Proteinase 3
MW in SDS-PAGE (3 isoforms)	26-28 kD	26.8-28.6 kD
Substrate specificity:		
α-naphthyl acetate	+	+
pH optimum	6.75-7.25	7.0
Hemoglobin	+	+
Elastin	+	+
Specific substrates for		
Elastase	−	−
Cathepsin G	−	−
Trypsin	−	−
Inhibitor profile:		
PMSF	+	+
Pepstatin	−	−
N-ethylmaleimide	−	−
1,10-Phenanthroline/EDTA	−	−

The data for the ACPA antigen are from Lüdemann et al. (1990) and for proteinase 3 from Kao et al. (1988).

INHIBITION OF THE ENZYMATIC ACTIVITY OF PROTEINASE 3 BY ACPA

The effect of ACPA on the enzymatic activity of proteinase 3 was analyzed in two different assays using the biologically relevant substrate elastin. For measuring elastinolysis in diffusion plates, agarose plates containing finely pulverized (smaller than 37 micron) bovine neck ligament elastin labeled with fluorescein (Elastin Products Company, Pacific, MO, USA) were prepared according to the manufacturer's instruction. To exclude interference caused by serum proteinase inhibitors (e.g. α-1-PI) IgG was purified from sera of a normal control and from ACPA-positive sera of various WG patients by protein G chromatography. The ACPA antigen or the control human leukocyte elastase (Calbiochem Corp., San Diego, CA, USA) were preincubated for 30 min with the IgG preparations and a buffer control and filled into the wells of a diffusion plate. After an incubation period of 16 hours at 37°C the plate was photographed with indirect illumination against a black background (Fig. 1). Elastinolysis in suspension was detected according to the manufacturer's instructions by measuring the release of fluorescein from the same elastin-fluorescein as described above, using a luminescence spectrometer (model LS 50; Perkin-Elmer Corp., Norwalk, CT, USA). Sample incubations were done as described above, except that release of fluorescein was already measured after an incubation period of 4 hours at 37°C.

Using the diffusion plate method, inhibition of elastinolysis can be calculated only roughly, since the border of the lysis zone is very diffuse (Fig. 1a), while human leukocyte elastase causes sharp lysis zones (Fig. 1b). But despite this limitation it is possible to see that IgG preparations from some sera inhibit elastinolysis to an extent that reflects the auto-antibody content (Table 4; samples 1-5). Other IgG preparations with high or moderate ACPA activity (Table 4; samples 6+7) induce only little or even no inhibition of elastinolysis, indicating that there are different types of antibodies. Some auto-antibodies seem to recognize an epitope at or near the active

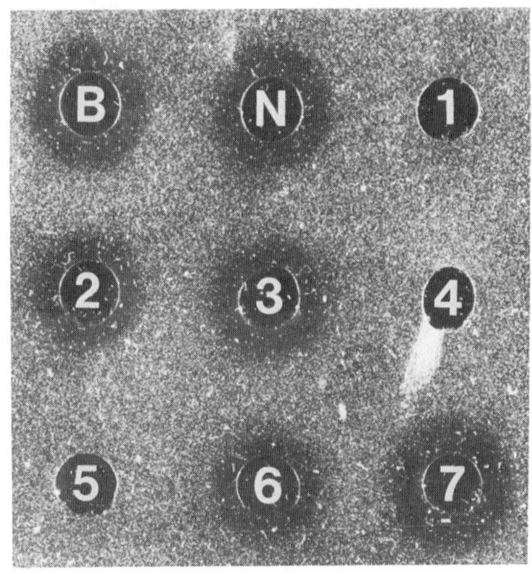 

a   b

Fig. 1. Inhibition of elastinolytic activity of the ACPA antigen by IgG preparations from ACPA-positive sera. The elastinolytic activity of the ACPA antigen was inhibited to different extents by ACPA-positive IgG preparations (a), whereas no inhibition of the control enzyme, human leukocyte elastase (b), was observed. Sample B represents the buffer control and the others are IgG preparations obtained by protein G chromatography from sera of a normal control (N) and of different patients with Wegener's granulomatosis (1-7). The ACPA content of the different samples and the calculated percentage of inhibition of elastinolytic activity are shown in Table 4.

site of the enzyme, while others are directed against different sites. The observed inhibition is not due to proteinase inhibitors that might contaminate the IgG preparations, as shown by incubating elastase with the same samples. Elastinolysis caused by elastase is not affected by the different IgG preparations (Fig. 1b), indicating that inhibition of proteinase 3 activity by ACPA really is a specific event.

In addition to using the diffusion plate method, elastinolysis was also measured in suspension. This method was employed to demonstrate that the observed inhibition of elastinolysis in the diffusion plate is really due to inhibition of enzymatic activity and not caused by inhibition of diffusion resulting from immunocomplex formation. Another advantage of this method is that elastinolysis can be quantitated more precisely by measuring the release of fluorescein. Although both methods yield quite comparable results (Table 4), it is remarkable that in suspension no total inhibition appeared.

PATHOLOGICAL CONSIDERATIONS

Proteinase 3 posesses elastinolytic activity and therefore may play a role in the destruction of tissue. It has been shown by Kao et al. (1988) that proteinase 3 is, along with elastase, only the second enzyme purified from human phagocytes that cau-

Table 4. Inhibition of Elastinolytic Activity of the ACPA
          Antigen by IgG Preparations from ACPA-Positive Sera

Sample[a]	ACPA Activity (U/ml)[b]	Inhibition of Elastinolysis (%) Diffusion Plate[c]	Suspension[d]
B	0	0	0
N	0	0	0
1	455	100	67
2	105	30	22
3	57	20	16
4	408	100	87
5	369	100	82
6	787	40	10
7	221	0	0

[a] Sample B represents the buffer control. The others are IgG preparations obtained by protein G chromatography from sera of a normal control (N) and of different patients with Wegener's granulomatosis (1-7).
[b] ACPA activity was measured using the ELISA method described by Lüdemann et al. (1988).
[c] Inhibition of elastinolysis in the diffusion plate was calculated by comparing the area of lysis caused by the buffer control with that of the test samples as shown in Figure 1.
[d] Inhibition of elastinolysis in suspension was calculated by comparing fluorescence values of the buffer control with those of the test samples.

ses experimental emphysema. Elastase has been reported to be partially responsible for tissue injury in vasculitis (Starkey, 1980) and to function as a major proteinase in granulomatous tissue remodeling (Izaki et al., 1986). Proteinase 3 also digests elastin and may therefore induce pathological alterations similar to elastase. In a preliminary experiment we found that binding of ACPA from serum of one WG patient to the antigen did not inhibit its enzymatic activity (Lüdemann et al., 1990). Therefore, we considered the hypothesis that the autoantibodies might prevent the enzyme from being inactivated by its natural inhibitors. A thus uninhibited proteinase digesting elastic fibers in vessel walls could cause a necrotizing vasculitis and may also be involved in granuloma formation, which are the two major histopathological findings in Wegener's granulomatosis. But this hypothesis must be questioned in view of the observation that autoantibodies of some patients do inhibit the enzymatic activity of proteinase 3 (Fig. 1; Table 4). Additionally, we performed experiments using the IgG preparations (Fig. 1; Table 4) and α-1-PI. After preincubating proteinase 3 with the IgG preparations, we added α-1-PI and measured elastinolysis in suspension. In all samples elastinolysis was inhibited to more than 95%, which was the same value that was obtained using α-1-PI together with the buffer control or the IgG preparation from a normal control. These results indicate that even those autoantibodies that do not inhibit proteinase 3 by themselves, are not able to prevent inhibition by α-1-PI. Therefore, the hypothesis outlined above is extremely unlikely to be true.

Inhibition of the enzymatic activity of proteinase 3 by the autoantibodies could of course also be associated with pathological alterations. It is known that enzyme inhibitors such as α-1-PI are inactivated by reactive oxygen species (Matheson et al., 1981), which are released by neutrophils. Therefore, it has been proposed that proteinase inhibitors are inactivated in the immediate vicinity of the neutrophils, resulting in protection of the serine proteinases from inhibition (Janoff, 1985; Carrell, 1986). One could speculate that autoantibodies might break through this protective shield and lead to inhibition of proteinase 3 in zones where the enzyme normally is active. In this way important biological functions of proteinase 3 might be blocked, but up to now hardly any of these functions are known. Only very recently Gabay et al. (1989) described an antimicrobial protein of human neutrophils that is most probably identical with proteinase 3 (Lüdemann et al., 1990). Therefore, further investigations into the biological functions of proteinase 3 and its physiological inhibitors are necessary to gain more insights into the pathomechanisms not only of vasculitis and pulmonary emphysema, but also of other inflammatory disorders.

Another hypothesis claiming that ACPA are a major pathological factor was raised recently by Jennette and Falk (1989). They found that in vitro both C-ANCA and P-ANCA are able to activate neutrophils, as evidenced by induction of a respiratory burst and degranulation. This neutrophil activation was most effective after priming of neutrophils with cytokines (e.g. tumor necrosis factor, TNF). They proposed that primed neutrophils express at their surfaces small amounts of primary granule constituents that are available to interact with the autoantibodies, resulting in full neutrophil activation. If ACPA cause neutrophil and monocyte activation in vivo within vessels a necrotizing inflammatory lesion would result. Indeed we measured high concentrations of TNF in serum of WG patients with active disease, while TNF levels were in the normal range in inactive phases of the disease (Rosenboom et al., 1989). These findings were confirmed by Deguchi et al. (1989) who found that transcription of the TNF-alpha gene is enhanced in WG. Moreover, we were able to show by immunoelectron microscopy that the ACPA antigen is expressed in small amounts on the plasma membrane of neutrophils and monocytes (Csernok et al., submitted). Therefore, this hypothesis is supported by several findings, and mechanisms like those outlined might in fact contribute to the pathogenesis of Wegener's granulomatosis.

REFERENCES

Baggiolini, M., Bretz, U., Dewald, B., and Feigenson, M. E., 1978, The polymorphonuclear leukocyte, _Agents Actions_, 8:3.

Carrell, R. W., 1986, α-1-antitrypsin: molecular pathology, leukocytes, and tissue damage, _J. Clin. Invest._, 78:1427.

Csernok, E., Lüdemann, J., Gross, W. L., and Bainton, D. F., Ultrastructural localization of proteinase 3, the target antigen of anti-cytoplasmic antibodies circulating in Wegener's granulomatosis, _J. Exp. Med._, submitted.

Deguchi, Y., Shibata, N., and Kishimoto, S., 1989, Enhanced transcription of TNF in systemic vasculitis, _Lancet_, ii:745.

Falk, R. J., and Jennette, J. C., 1988, Anti-neutrophil cytoplasmic autoantibodies with specificity for myeloperoxidase in patients with systemic vasculitis and idiopathic necrotizing and crescentic glomerulonephritis, <u>N. Engl. J. Med.</u>, 318:1651.

Gabay, J. E., Scott, R. W., Campanelli, D., Griffith, J., Wilde, C., Marra, M. N., Seeger, M., and Nathan, C. F., 1989, Antibiotic proteins of human polymorphonuclear leukocytes, <u>Proc. Natl. Acad. Sci. USA</u>, 86:5610.

Goldschmeding, R., Tetteroo, P. A. T., von dem Borne, A. E. G. Kr., and Kallenberg, C. G. M., 1987, Anti-neutrophil cytoplasm antibodies in Wegener's granulomatosis are not directed against alkaline phosphatase, <u>Lancet</u>, i:1489.

Goldschmeding, R., Cohen Tervaert, J. W., van der Schoot, C. E., van der Veen, C., Kallenberg, C. G. M., and von dem Borne, A. E. G. Kr., 1989a, ANCA, anti-myeloperoxidase, and anti-elastase: three members of a novel class of autoantibodies against myeloid lysosomal enzymes, <u>Acta Pathol. Microbiol. Immunol. Scand.</u>, 97(suppl. 6):48.

Goldschmeding, R., Ten Bokkel Huinink, D., Faber, N., Tetteroo, P. A. T., Vroom, T. M., Hack, C. E., and von dem Borne, A. E. G. Kr., 1989b, Identification of the ANCA-antigen as a novel myeloid lysosomal serine protease, <u>Acta Pathol. Microbiol. Immunol. Scand.</u>, 97(suppl. 6):46.

Gross, W. L., Lüdemann, G., Kiefer, G., and Lehmann, H., 1986, Anti-cytoplasmic antibodies in Wegener's granulomatosis, <u>Lancet</u>, i:806.

Gross, W. L., Lüdemann, J., and Schröder, J.-M., 1987, Anti-neutrophil cytoplasm antibodies in Wegener's granulomatosis are not directed against alkaline phosphatase, <u>Lancet</u>, i:1488.

Gross, W. L., 1989, Wegener's granulomatosis: new aspects of the disease course, immunodiagnostic procedures and stage-adapted treatment, <u>Sarcoidosis</u>, 6:15.

Heck, L. W., Rostand, K. S., Hunter, F. A., and Bhown, A., Isolation, characterization, and amino-terminal amino acid sequence analysis of human neutrophil cathepsin G from normal donors, <u>Anal. Biochem.</u>, 158:217.

Izaki, S., Okamoto, M., Hsu, P.-S., Epstein, W. L., and Fukuyama, K., 1986, Characterization of elastase associated with granulomatous tissue remodeling, <u>J. Cell. Biochem.</u>, 32:79.

Janoff, A., 1985, Elastase in tissue injury, <u>Ann. Rev. Med.</u>, 36:207.

Jennette, J. C., and Falk, R. J., 1988, Anti-neutrophil cytoplasmic autoantibodies, <u>N. Engl. J. Med.</u>, 319:1417.

Jennette, J. C., and Falk, R. J., 1989, Anti-neutrophil cytoplasmic autoantibodies: new insight into crescentic glomerulonephritis, pulmonary-renal syndrome and systemic vasculitis, <u>AKF Nephr. Letter</u>, 6:11.

Kao, R. C., Wehner, N. G., Skubitz, K. M., Gray, B. H., and Hoidal, J. R., 1988, Proteinase 3: a distinct human polymorphonuclear leukocyte proteinase that produces emphysema in hamsters, <u>J. Clin. Invest.</u>, 82:1963.

Lockwood, C. M., Bakes, D., Jones, S., Whitaker, K. B., Moss, D. W., and Savage, C. O. S., 1987, Association of alkaline phosphatase with an autoantigen recognized by circulating anti-neutrophil antibodies in systemic vasculitis, <u>Lancet</u>, i:716.

Lüdemann, G., and Gross, W. L., 1987, Autoantibodies against cytoplasmic structures of neutrophil granulocytes in Wegener's granulomatosis, <u>Clin. Exp. Immunol.</u>, 69:350.

Lüdemann, G., Nölle, B., Rautmann, A., Rosenboom, S., Kekow, J., and Gross, W. L., 1988a, Antizytoplasmatische Antikörper als Seromarker und Aktivitätsparameter der Wegener'schen Granulomatose: Eine prospektive Studie, <u>Dtsch. Med. Wochenschr.</u>, 113:413.

Lüdemann, J., Utecht, B., and Gross, W. L., 1988b, Detection and quantification of anti-neutrophil cytoplasm antibodies in Wegener's granulomatosis by ELISA using affinity-purified antigen, <u>J. Immunol. Methods</u>, 114:167.

Lüdemann, J., Utecht, B., and Gross, W. L., 1990, Anti-neutrophil cytoplasm antibodies in Wegener's granulomatosis recognize an elastinolytic enzyme, <u>J. Exp. Med.</u>, 171: in press.

Matheson, N. R., Wong, P. S., Schuyler, M., and Travis, J., 1981, Interaction of human α-1-proteinase inhibitor with neutrophil myeloperoxidase, <u>Biochemisty</u>, 20:331.

Nölle, B., Specks, U., Lüdemann, J., Rohrbach, M. S., DeRemee, R. A., and Gross, W. L., 1989, Anticytoplasmic autoantibodies: their immunodiagnostic value in Wegener granulomatosis, <u>Ann. Intern. Med.</u>, 111:28.

Matsudaira, P., 1987, Sequence from picomole quantities of proteins electroblotted onto polyvinylidene difluoride membranes, <u>J. Biol. Chem.</u>, 262:10035.

Parlevliet, K. J., Henzen-Logmans, S. C., Oe, P. L., Bronsveld, W., Balm, A. J. M., and Donker, A. J. M., 1988, Antibodies to components of neutrophil cytoplasm: a new diagnostic tool in patients with Wegener's granulomatosis and systemic vasculitis, <u>Q. J. Med.</u>, 66:55.

Rasmussen, N., Borregaard, N., and Wiik, A., 1987, Anti-neutrophil cytoplasm antibodies in Wegener's granulomatosis are not directed against alkaline phosphatase, <u>Lancet</u>, i:1488.

Rasmussen, N., Wiik, A., Høier-Madsen, M., Borregaard, N., and van der Woude, F., 1988, Anti-neutrophil cytoplasm antibodies 1988, <u>Lancet</u>, i:706.

Rosenboom, S., Rautmann, A., and Gross, W.L., 1989, Tumor necrosis factor in Wegener's granulomatosis, <u>in</u>: Abstract book of the 7th International Congress of Immunology, Berlin, Gustav Fischer, Stuttgart, New York.

Savage, C. O. S., Winearls, C. G., Jones, S., Marshall, P. D., and Lockwood, C. M., 1987, Prospective study of radioimmunoassay for antibodies against neutrophil cytoplasm in diagnosis of systemic vasculitis, <u>Lancet</u>, i:1389.

Sinha, S., Watorek, W., Karr, S., Giles, J., Bode, W., and Travis, J., 1987, Primary structure of human neutrophil elastase, <u>Proc. Natl. Acad. Sci. USA</u>, 84:2228.

Specks, U., Wheatley, C. L., McDonald, T. J., Rohrbach, M. S., and DeRemee, R. A., 1989, Anticytoplasmic autoantibodies in the diagnosis and follow-up of Wegener's granulomatosis, <u>Mayo Clin. Proc.</u>, 64:28.

Starkey, P. M., 1980, The role of cellular elastases in inflammation, <u>Front. Matrix Biol.</u>, 8:188.

Thompson, R. A., and Lee, S. S., 1989, Antineutrophil cytoplasmic antibodies, <u>Lancet</u>, i:670.

Van der Woude F. J., Rasmussen N., Lobatto S., Wiik, A., Permin, H., van Es, L. A., van der Giessen, M., van der Hem, G. K., and The, T. H., 1985, Autoantibodies against neutrophils and monocytes: tool for diagnosis and marker of disease activity in Wegener's granulomatosis, <u>Lancet</u>, i:425.

Wehner, N. G., 1987, Biochemistry and function of proteases isolated from the cytoplasmic granules of polymorphonuclear leukocytes, PH.D. Thesis, University of Minnesota.

NEUTROPHIL CARBOHYDRATE METABOLISM IN PATIENTS WITH ESSENTIAL HYPERTENSION AND UREMIA

M. Haag-Weber, P. Schollmeyer, W. H. Hörl

Department of Medicine, Division of Nephrology
University of Freiburg, FRG

INTRODUCTION

Infectious complications result in significant morbidity
and mortality of patients with end-stage renal disease. Uremia
is an immunocompromised state due to direct effects of uremic
toxins and indirect factors, e. g. malnutrition, dialysis membranes with their effects on complement system and white blood
cells, or vascular access for dialysis providing a portal of
entry for microorganisms (1). Dysfunction of polymorphonuclear
(PMN) cells in uremia includes adherence, the first step in
neutrophil migration, chemotaxis, phagocytotic capacity, generation of reactive oxygen intermediates or intracellular killing of bacteria (2).

A reduction in generation of chemotactic activity in ex-
perimentally induced acute renal failure has been reported by
Clark et al. (3). This reduction was proportional to the
degree of azotemia. Baum et al. (4) demonstrated that chemotaxis of PMN was impaired in the presence of undialysed uremic
sera but that hemodialysis seemed to correct the defect. A
chemotactic inhibitor was found in sera of uremic patients
acting directly on leukotactic factors (C_3 and C_5 chemotactic
fragment and bacterial factor) to render them irreversibly
inactive (5). Pedersen et al. (6) studied the ability of serum
to attract PMN in uremic patients before and after hemodialysis. The chemotactic responses towards serum from uremic
patients were significantly decreased prior to hemodialysis
and normalized thereafter.

Phagocytosis in uremic patients is either normal or
slightly diminished. Hirabayashi et al. (7) observed decreased
phagocytic uptake of IgG-coated particles in patients before
hemodialysis therapy, which was restored by hemodialysis. Phagocytotic responsiveness of PMN increased after exposure to
cuprophane filter although exposure to polyacrylonitrile and
polysulfone hemodialyzers did not affect phagocytotic activity
(8). A dramatic fall in phagocytotic activity after 15 minutes

of hemodialysis with cuprophane membranes was observed. Reused
cuprophane caused a minor decrease in phagocytotic activity,
whereas hemodialysis with polyacrylonitrile and polysulfone
produced no significant change in their populations at this
time point (8).

Kinetic measurements of the serum-independent uptake of
IgG-coated polyvinyl toluene latex particles by isolated PMN
cells have been performed in patients undergoing hemodialysis
or peritoneal dialysis (9). The mean phagocytic rate for the
patient group was significantly reduced when compared to the
reference group of apparently healthy individuals. After 4
months of adequate dialysis, the phagocytic uptake was
significantly higher than the uptake shortly after the
beginning of dialysis. It was suggested that elevated serum
phosphate levels may in part be responsible for the impaired
phagocytic activity of uremic patients (9).

OXIDATIVE METABOLISM AND GLUCOSE METABOLISM IN UREMIA

Ritchey et al. (10) studied superoxide anion production
and luminol-amplified chemiluminescence in PMN from chronic
hemodialysis patients and in age-matched controls in the resting state and response to phorbol myristate acetate (PMA).
Studies in autologous serum showed higher chemiluminescence
resting values in PMN from hemodialysis patients with a significant reduction after dialysis. Crossincubation studies indicated that this is a result of factor(s) in the patients' serum. In response to PMA, neutrophils from chronic hemodialysis
patients in autologous serum had significantly less of an increase in chemiluminescence as compared with controls, suggesting that there is a defect intrinsic to the patient PMN
(10). Hirabayashi et al. (7) also found impaired hydrogen
peroxide production by PMA-stimulated PMN before dialysis
which was restored to the control level by hemodialysis.

Different changes in oxidative metabolism have been
observed in PMN when exposed to a cuprophane filter but not
with polyacrylonitrile filter membrane (11). There was a
reduced PMN activity of both chemiluminescence and H_2O_2 as a
consequence of hemodialysis when a cuprophane dialyzer was
employed; however, the diminished activity was no longer
detectable after the filter had been hemophane modified (12).

Markert et al. (13) found an increase of phorbol myristate acetate-stimulated lucigenin-amplified chemiluminescence
of PMN isolated 5 and 60 minutes during hemodialysis with
cuprophane, cellulose acetate and polyacrylonitrile during the
first and/or second use as compared to the control cells.
Luminol increased chemiluminescence during the initial use of
three filters: cuprophane, polycarbonate and polysulfone.

Although numerous studies have affirmed changes of PMN
metabolism and function in uremia, most of the evidence has
come from data obtained before and during hemodialysis.
However, information comparing PMN function after hemodialysis
treatment with that prior to treatment are lacking. Studies on
glucose metabolism have revealed that glucose intolerance and
impaired utilization of glucose are improved significantly

after uremic patients have undergone hemodialysis treatment. This suggests that a circulating plasma factor is responsible for inducing insulin resistance (14,15,16). A variety of normal tissues obtained from animals have inhibited basal uptake of glucose following incubation with whole or partially purified sera obtained from uremic patients (17,18,19). The insulin stimulation of glucose uptake and metabolism of rat adipocytes also was reduced following preincubation with serum from uremic patients (20). This irregularity occurs without affecting either insulin binding or the antilipolytic action of insulin (15).

Cellular activation of PMN by microbial antigens and certain toxins are achieved by the stimulation of phosphoinositol (PI) turnover and activation of protein kinase C. For example, the stimulus-response coupling in PMN is induced by a chemotactic peptide. Concanavalin A, immune complexes, and PMA showed an increase of calcium uptake proceeding onset of degranulation and of O_2^--generation (21). The stimulation of hexose transport by PMNs is associated with the activation of protein kinase C (22). Furthermore, pertussis toxin that is known to involve the activation of G proteins effectively stimulates PI turnover with regard to increased formation of IP_3 and an increased elevation of intracellular calcium (23).

We have observed that PMN from uremic patients have diminished responsiveness to the chemotactic peptide FMLP-induced hexose uptake, enzyme release, and chemotaxis (24). This lack of responsiveness is improved after the patients have undergone hemodialysis. Moreover, when normal PMN are exposed to various concentrations of ultrafiltrates from uremic patients, there was an increased loss of responsiveness of these cells with regard to hexose uptake (25) and enzyme release. In addition , washing the PMN with buffered saline restored their function to normal which suggests that a plasma-derived constituent is responsible for these effects. There is further evidence for the existence of such a putative entity from initial experiments that have employed one species with a 8-30 Kd molecular weight (MW) that was partly purified by column chromatography (Sephadex G-100) from the ultrafiltrate of uremic patients. The factor greater than 10 Kd MW appears to be heat lable which suggests that it can be further purified, characterized and its biochemical and immunochemical properties further elucidated.

Recently, a novel protein (molecular weight of 28 Kd) was isolated and characterized from uremic serum of patients undergoing regular hemodialysis therapy. This polypeptide inhibits the uptake of glucose, chemotaxis, oxidative metabolism, and intracellular bacterial killing by polymorphonuclear leukocytes (26). The IC_{50} of the granulocyte inhibiting protein required to inhibit both the biochemical and functional changes is in the nanomolar range. Therefore, the efficacy of the protein is well within the range of physiological effector substances (table 1). A specific rabbit polyclonal antibody raised against the protein nullified these inhibitory changes (26).

In summary, a granulocyte inhibitory protein (GIP) was isolated from ultrafiltrates of patients on regular hemodialy-

Table 1. Biological activity of granulocte inhibiting protein

Test condition	GIP concentration which inhibits 50% of maximal stimulation (IC_{50})
^{3}H d-glucose uptake[*]	5.2 ± 1.2 μg/ml (n=5)
Chemotaxis[*]	7.2 ± 1.7 μg/ml (n=6)
Oxidative metabolism[*]	8.1 ± 1.4 μg/ml (n=5)

[*] stimulation with FMLP; mean ± S.E.M.

sis treatment using the polysulfone membrane (Fresenius, Oberursel, FRG). The protein is capable of inhibiting four fundamental PMN cell function. Its exact role remains to be elucidated.

GLYCOGEN METABOLISM IN DIALYSIS-PATIENTS

Dialysis and different membrane materials effect glucose uptake and glycogen metabolism of PMNs. Both phagocytotic capacity and carbohydrate metabolism of PMNs are impaired in chronically uremic patients. Further alterations may occur during hemodialysis treatment. Studies of Ohlsson and Olsson (27) demonstrated that granulocyte myeloperoxidase, lactoferrin, collagenase and elastase are released simultaneously following phagocytosis. Plasma levels of these main granulocyte components also undergo progressive elevation during hemodialysis (28,29,30).

Recent studies showed that the activity of the glycogen-degrading enzyme phosphorylase $\underline{a}$ increases significantly during hemodialysis with polymethylmethacrylate (PMMA) but not with dialyzers made of polysulfone. Predialysis phosphorylase $\underline{a}$ activity in the PMN of hemodialysis patients were comparable to that of control subjects. The activity of the active I-form of glycogen synthetase was significantly lower in PMNs of hemodialysis patients compared with healthy controls. The enzyme activity decreased significantly in patients dialyzed with PMMA dialyzers but increased significantly in PMNs of patients dialyzed with dialyzers made of polysulfone.

Glycogen metabolism in PMNs of CAPD-patients is in contrast to hemodialysis patients not affected by PMN-activation by different membrane materials. Figure 1 and table 2 summarizes our experiments with PMNs from CAPD-patients. Both total glycogen synthetase activity (D- and I-forms) and the active I-form of glycogen synthetase as well as glycogen content of CAPD patients were significantly lower than in PMNs of healthy controls. Again, phosphorylase $\underline{a}$ activity did not differ from enzyme activity measured in PMNs of hemodialysis patients and healthy subjects. However, glucose uptake in PMNs of CAPD patients was higher than in hemodialysis patients even at the end of dialysis therapy but not in the range of healthy subjects.

In conclusion, impaired PMN glycogen synthesis and
glucose uptake in uremia may cause or contribute to granulocyte dysfunction described in patients with renal insufficien-
cy. Enhanced susceptibility to infections persists also in HD
and CAPD patients probably due to the fact that both treatment
modalities do not restore uremia induced abnormalities of PMN
carbohydrate metabolism and other PMN disturbances.

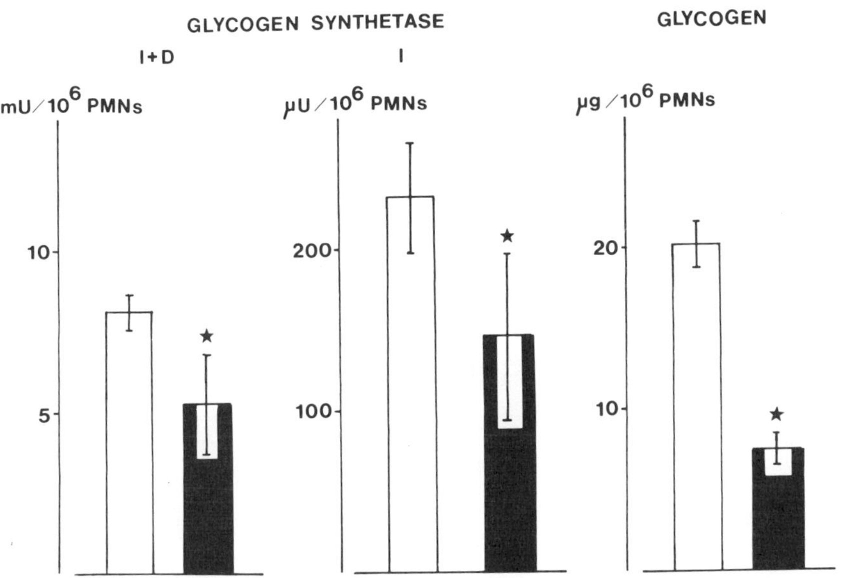

Figure 1. Total activity of glycogen synthetase (I+D-form), of
active I-form of glycogen synthetase and glycogen
content in PMNs of healthy controls (□) and of CAPD
patients (■)

CARBOHYDRATE METABOLISM IN PMN OF PATIENTS WITH ESSENTIAL HYPERTENSION (EH)

High blood pressure is associated with glucose intolerance or non-insulin dependent diabetes mellitus (31).
Preliminary evidence has been presented that essential hypertension is an insulin resistant state involving glucose but not
lipid or potassium metabolism. It has been shown that insulin-
induced stimulation of whole body glucose uptake is markedly
reduced in patients with essential hypertension (32). Insulin,
however, does not affect glucose uptake by PMNs in the basal

Table 2. ^{3}H-d-glucose uptake in PMNs of healthy controls (Co), CAPD-patients, hemodialysis patients (RDT) before and after hemodialysis with polysulfone and PMNA

	$\Delta\,^3$H-d-glucose uptake (cpm/2×10^5 PMNs)
Co	7,794 ± 1,641
CAPD	6,018 ± 1,603
RDT (Polysulfone) before	2,256 ± 923*§
RDT (Polysulfone) after	4,410 ± 615*
RDT-PMMA before	3,711 ± 1,189*§
RDT-PMMA after	4,890 ± 1,671*

Mean values ± SEM
*p <0.05 Co vs CAPD, RDT-patients
§p <0.05 CAPD vs RDT-patients

Table 3. Active I-form of glycogen synthetase and total enzyme activity (D+I-form) of PMNs of controls and EH-patients treated with and without calcium channel blockers

	Glycogen synthetase	
	I-form μU/10^6PMNs	D+I-form mU/10^6PMNs
Controls (n=15)	246 ± 34.9	6.23 ± 0.52
EH-patients (n=12) with calcium channel blockers	156 ± 15.9*	6.30 ± 0.64
EH-patients (n=9) without calcium channel blockers	184 ± 38.0	7.37 ± 0.87

Mean values ± SEM
*p <0.05 EH-patients versus controls

state and after stimulation with chemotactic peptides (22).
PMN glucose uptake is inhibited in uremic patients (33), and
uremia is also associated with an insulin-resistant state
(34). Endocrine and metabolic abnormalities have been reported
in hypertensive patients. Significantly elevated PTH and 1,25
dihydroxyvitamin D as well as decreased serum calcitonin
levels were observed in patients with low renin hypertension
but not in hyertensive patients with high plasma renin
activity (35). Significantly higher serum PTH values were also
observed in our EH-patient group. In the fasting, postab-
sorptive state, venous plasma insulin and glucose values were
comparable in untreated patients with essential hypertension
and healthy subjects. After glucose load, however, plasma
insulin and glucose values were significantly higher in
hypertensives (32). Our group of non-fasting EH-patients
displayed also significantly higher blood glucose values and a
tendency of increased inuslin levels. Our data show impaired
glucose uptake in PMNs of EH-patients demonstrating a defect
in glucose uptake in hypertensives independent of insulin.
Table 3 depicts total activity of glycogen synthetase (I+D-
form) and the active I-form of the enzyme in PMNs of EH-
patients and controls. EH-patients can be subdivided into two

Table 4. In vitro effects of diltiazem and verapamil on O_2-
production of PMNs obtained from healthy subjects

	PMN O_2- production (nmol/min/10^6PMNs)		
	without FMLP	with FMLP	difference
PMNs in buffer	1.40 ± 0.19	2.90 ± 0.30	1.50 ± 0.23
PMNs in diltiazem (0.5 μg/ml)	1.52 ± 0.26	2.40 ± 0.43	0.90 ± 0.32[*]
PMNs in verapamil (0.25 μg/ml)	1.35 ± 0.17	2.04 ± 0.33[*]	0.60 ± 0.37[*]

Mean values ± SEM from 12 experiments
[*]p <0.05 PBS versus calcium channel blockers

groups. Active I-form of glycogen synthetase was significantly
reduced in EH-patients under treatment with calcium channel
blockers, whereas enzyme activity was not different between
EH-patients treated with a variety of antihypertensive agents
and controls (table 3). Glycogen content of PMNs isolated from
healthy controls and EH-patients did not differ significantly.
EH-patients treated with calcium channel blockers showed lower
PMN glucose uptake than healthy subjects or EH-patients under
antihypertensive therapy with converting enzyme inhibitors or
beta receptor blocking agents. In vitro, both verapamil and

diltiazem decreased PMN glucose uptake of healthy subjects.
One has to keep in mind, however, that the concentrations used
here are about 50 times greater than pharmacologically active
blood concentrations of the drug. An increase of intracellular calcium enhances phagocytotic capacity of PMNs and also
the release of neutrophil enzymes (36), which can be reduced
by calcium channel blockers (37,38). Intact hexose monophosphate shunt activity is necessary for sustained normal
phagocyte function (because the oxidase is NAPDH dependent).
In parallel with studies of carbohydrate metabolism we
examined a further parameter important for normal PMN function. We could clearly demonstrate that pharmacological doses
of calcium channel blockers reduce significantly O_2^--produc-
tion of PMNs of healthy subjects in vitro (table 4). Whether
the impaired carbohydrate metabolism and O_2^--release of PMNs of
EH-patients caused by calcium channel blockers is of clinical
importance needs to be determined.

REFERENCES

1. Tolkoff, N.E., Rubin, R.H.: Uremia and host defenses. <u>N.
    Engl. J. Med.</u> 322:770-772 (1990).
2. Haag-Weber, M., Hable, M., Schollmeyer, P., Hörl, W.H.:
    Metabolic response of neutrophils to uremia and dialysis.
    <u>Kidney Int.</u> 36 (Suppl 27):293-298 (1989).
3. Clark, R.A., Hamony, B.H., Ford, G., Kimball, H.R.:
    Chemotaxis in acute renal failure. <u>J. Infect. Dis.</u> 126:
    460-463 (1972).
4. Baum, J., Cestero, R.V.M., Freeman, R.B.: Chemotaxis of
    polymorphonuclear leukocytes and delayed hypersensitivity
    in uremia. <u>Kidney Int.</u> 7 (Suppl 2):147-153 (1975).
5. Siriwatratananonta, P., Sinsakul, V., Stern, K., Slavin,
    R.G.: Defective chemotaxis in uremia. <u>J. Lab. Clin. Med.</u>
    92:402-407 (1978).
6. Pedersen, J.O., Knudsen, F., Nielsen, A.H., Grunnet, N.:
    The ability of uremic serum to induce neutrophil chemo-
    taxis in relation to hemodialysis 5:24-28 (1987).
7. Hirabayashi, Y., Kobayashi, T., Nishikawa, A., Aoki, T.,
    Takaya, J., Kobayashi, Y.: Oxidative metabolism and
    phagocytosis of polymorphonuclear leukocytes in patients
    with chronic renal failure. <u>Nephron</u> 49:305-312 (1988).
8. Vanholder, R.C., Dhondt, A., Ringoir, S.M.G.: Challenge
    of phagocyte metabolism by extracorporeal test. <u>Trans.
    Am. Soc. Artif. Intern. Organs</u> 34:214-218 (1988).
9. Hällgren, R., Fjellström, K.E., Venge. P.: Kinetic
    studies of phagocytosis. II. The serum-independent uptake
    of IgG-coated particles by polymorphonuclear leukocytes
    from uremic patients on regular dialysis treatment. <u>J.
    Lab. Clin. Med.</u> 94:277-284 (1979).
10. Ritchey, E.E., Wallin, J.D., Shah, S.V.: Chemilumi-
    nescence and superoxide anion production by leukocytes
    from chronic hemodialysis patients. <u>Kidney Int.</u> 19:349-
    358 (1981).
11. Nguyen, A.T., Lethias, C., Zingraff, J., Herbelin, A.,
    Naret, C., Descamps-Latscha, B.: Hemodialysis membrane-
    induced activation of phagocyte oxidative metabolism de-
    tected in vivo and in vitro within microamounts of whole
    blood. <u>Kidney Int.</u> 28:158-167 (1985).

12. Kolb, G., Schönemann, H., Fischer, W., Bittner, K., Lange, H., Höffken, H., Damann, V., Joseph, K., Havemann, K.: Hemodialysis with cuprophane membranes leads to alteration of granulocyte oxidative metabolism and leukocyte sequestion in the lung. In: Hörl, W.H., Heidland, A. (eds) Proteases: Potential Role in Health and Disease II. Plenum, New York, pp. 377-384 (1988).

13. Markert, M., Heierli, C., Kuwahara, T., Frei, J., Wauters, J.P.: Dialyzed polymorphonuclear neutrophil oxidative metabolism during dialysis: a comparative study with 5 new and reused membranes. Clin. Nephrol. 29:129-136 (1988).

14. McCaleb, M.L., Izzo, M.S., Lockwood, D.H.: Characterization and partial purification of a factor from uremic human serum that induces insulin resistance. J. Clin. Invest. 75:391-396 (1985).

15. DeFronzo, Tobin, J.D., Rowe, J.W., Andres, R.: Glucose intolerance in uremia. J. Clin. Invest. 62:425-435 (1978).

16. Hampers, C.L., Soeldner, J.S., Doak, P.B., Merrill, J.P.: Effect of chronic renal failure and hemodialysis on carbohydrate metabolism. J. Clin. Invest. 45:1719-1731 (1966).

17. Balesteri, P., Rindi, P., Biagini, M., Giovanetti, S.: Effects of uraemic serum, urea, creatinine and methyl-guanidine on glucose metabolism. Clin. Sci. 42:395-404 (1972).

18. Morgan, J.M., Morgan, R.E.: Study of the effect of uremic metabolites on erythrocyte glycosis. Metabolism 13:629-635 (1964).

19. Dzurik, R.: Metabolic alterations caused by uremia. Proc. Eur. Dial. Tranplant. Assoc. 17:577-586 (1980).

20. McCaleb, M.L., Mevorach, R., Freeman, R.B., Izzo, M.S., Lockwood, D.H.: Induction of insulin resistance in normal adipose tissue by uremic human serum. Kidney Int. 25:416-421 (1984).

21. Korchak, H.M., Rutherford, L.E., Weissmann, G.: Stimulus response coupling in the human neutrophiles. I. Kinetic analysis of changes in calcium permeability. J. Biol. Chem. 259:4070 (1984).

22. McCall, C., Schmitt, J., Cousart, S., O'Flaherty, J., Bass, D., Wykle, R.: Stimulation of hexose transport by human polymorphonuclear leucocytes: a possible role of protein kinase C. Biochem. Biophys. Res. Comm. 126:450-456 (1985).

23. Krause, K.H., Schlegel, W., Wollheim, C.B., Andersson, T., Waldvogel, F.A., Lew, P.D.: Chemotactic peptide activation of human neutrophils and HL-60 cells. Pertussis toxin reveals correlation between inositol triphosphate generation, calcium ion transients, and cellular activation. J. Clin. Invest. 76:1348-1354 (1985).

24. Haag-Weber, M., Hable, M., Schollmeyer, P., Hörl, W.H.: Hemodialysis improves carbohydrate metabolism in polymorphonuclear neutrophils (PMN) (abstract). Kidney Int. 35:248 (1989).

25. Haag-Weber, M., Schollmeyer, P., Hörl, W.H.: Neutrophil activation during hemodialysis. In: Hörl, W.H., Schollmeyer, P.J. (eds) New Perspectives in Hemodialysis, Peritoneal Dialysis, Arteriovenous Hemofiltration, and Plasmapheresis. Plenum, New York, pp. 27-37 (1989).

26. Hörl, W.H., Haag-Weber, M., Georgopoulos, A., Block, L.H.: The physicochemical charaterization of a polypeptide present in uremic serum that inhibits the biological activity of polymorphonuclear cells. Proc. Natl. Acad. Sci. USA 87:6353-6357 (1990).

27. Ohlsson, K., Olsson, I.: Neutral proteases of human granulocytes. III. Interaction between human granulocyte elastase and plasma protease inhibitors. Scand. J. Lab. Invest. 34:349-355 (1974).

28. Hörl, W.H., Steinhauer, H.B., Schollmeyer, P.: Plasma levels of granulocyte elastase during hemodialyis: Effects of different dialyzer membranes. Kidney Int. 28: 791-796 (1985).

29. Hörl, W.H., Jochum, M., Heidland, A., Fritz, H.: Release of granulocyte proteinases during hemodialysis. Am. J. Nephrol. 3:213-217 (1983).

30. Hörl, W.H., Schäfer, R.M., Heidland, A.: Effect of different dialyzers on proteinase inhibitors during hemodialysis. Am. J. Nephrol. 5:320-326 (1985).

31. Modan, M., Halkin, H., Almong, S., Lusky, A., Eshkol, A., Shefi, M., Shifrit, H., Fuchs, Z.: Hyperinsulinemia: a link between hypertension obesity and glucose intolerance. J. Clin. Invest. 75:809-817 (1985).

32. Ferrannini, E., Buzzigoli, G., Bonadonna, R., Giorico, A., Oleggini, M., Graziadei, L., Pedrinelli, R., Brandi, L., Bevilacqua, S.: Insulin resistance in essential hypertension. N. Engl. J. Med. 317:350-357 (1987).

33. Briggs, W.A., Sillix, D.H., Mahajan, S., McDonald, F.D.: Leukocyte metabolism and infection in uremia. Kidney Int. 24 (Suppl 16):93-96 (1983).

34. DeFronzo, R.A., Smith, D., Alvestrand, A.: Insulin action in uremia. Kidney Int. 24 (Suppl 16):102-114 (1983).

35. Resnick, L.M., Laragh, J.H: Renin, calium metabolism and the pathophysiologic basis of antihypertensive therapy. Am. J. Cardiol. 56:68H-74H (1985).

36. Lew, D.P.: Receptor signalling and intracellular calcium in neutrophil activation. Europ. J. Clin. Invest. 19: 338-346 (1989).

37. Haag-Weber, M., Schollmeyer, P., Hörl, W.H.: Granulocyte activation in the absence of complement activation: Inhibition by calcium channel blockers. Europ. J. Clin. Invest. 18:380-385 (1988).

38. Riegel, W., Spillner, G., Schlosser, V., Hörl, W.H.: Plasma levels of main granulocyte components during cardiopulmonary bypass. J. Thorac Cardiovasc Surg 95: 1014-1019 (1988).

EFFECT OF SULFATED GLYCOSAMINOGLYCANS ON

THE INHIBITION OF NEUTROPHIL ELASTASE BY α_1-PROTEINASE INHIBITOR

Klaus Frommherz and Joseph G. Bieth

INSERM U 237, Faculté de Pharmacie

74 route du Rhin, F - 67400 Illkirch, France

INTRODUCTION

The azurophil granules of neutrophils contain a number of proteolytic enzymes including elastase, a 30 kDa serine proteinase. This cationic glycoprotein does not only solubilize elastin but also cleaves a number of extracellular matrix and plasma proteins (e.g. type IV collagen, immunoglobulins) [1].

The physiologic functions of neutrophil elastase include proteolysis of phagocytosed proteins, neutrophil migration and tissue remodeling following injury [2]. The extremely high elastase content of neutrophils (the granule's elastase concentration is 9 mM [3]) requires potent control mechanisms to prevent undesirable extracellular protein degradation during neutrophil activation.

α_1-proteinase inhibitor is thought to be the most important component of the extracellular antielastase screen. Its genetic deficiency indeed leads to elastase-mediated lung emphysema. It is a 53 kDa glycoprotein which inhibits serine proteinases irreversibly by forming a denaturant-stable complex with them. Among all proteinases tested, neutrophil elastase is inhibited with the highest rate [4]. This fast-acting inhibition (k_{ass} = 1.3×10^7 M^{-1} s^{-1}), together with the high extracellular concentration of the inhibitor (plasma conc ~ 30 µM) renders the inhibition process extremely efficient : dt, the delay time of inhibition of elastase in plasma, i.e. the time required to get almost full inhibition of the enzyme[5], is 13 milliseconds.

Despite this high efficacy, situations are known where this inhibition process fails : (i) α_1-proteinase inhibitor may be inactivated by metallo or cysteine proteinases that are not inhibited by it but use it as a substrate, (ii) Met 358, the P_1 residue of the active center of the inhibitor may be oxidized into methionine sulfoxide by the myeloperoxidase + $H_2 O_2$ + halid system of the neutrophil or by other oxidants including those present in cigarette smoke ; as a consequence, k_{ass} decreases 2000-fold so that the protein becomes a poorly efficient elastase inhibitor, (iii) due to its high intragranular concentration (*vide supra*), the elastase

New Aspects of Human Polymorphonuclear Leukocytes
Edited by W.H. Hörl and P.J. Schollmeyer, Plenum Press, New York, 1991

concentration may exceed that of α_1-proteinase inhibitor in the immediate vicinity of an activated or a "dying" neutrophil, (iv) the close contact between a neutrophil and an extracellular matrix protein may hinder the access of α_1-proteinase inhibitor and thus favor proteolysis. Here we report on the failure of α_1-proteinase inhibitor to efficiently inhibit neutrophil elastase in the presence of heparin and other sulfated glycosoaminoglycans.

Heparin, a naturally occurring sulfated polysaccharide, is a mixture of polymers composed of disaccharide units with three sulfated groups per unit. The disaccharide units are formed of iduronic acid with one 0-sulfated group and glucosamine with an 0-and a N-sulfated group. Some polysaccharide chains also contain domains in which glucosamine contains an additional 0-sulfated group. Such domains are believed to react with antithrombin III with resultant acceleration of the rate of thrombin inhibition [6]. Heparin has been shown to bind neutrophil elastase through electrostatic interactions and to partially inhibit the activity of the enzyme [7].

RESULTS

We have measured k_{ass}, the association rate constant for the inhibition of elastase by α_1-proteinase inhibitor [4] in the absence and presence of commercial high molecular weight heparin (Mr = 13.5-15 kDa) and low molecular weight heparin (Mr = 3.7 kDa). The rate constant decreased with the concentration of both heparins and then reached a plateau. The maximal decreases in k_{ass} are reported in table 1 together with the maximal effects produced by other glycosaminoglycans.

Table 1. Effect of glycosaminoglycans on k_{ass}, the rate constant for the inhibition of neutrophil elastase by α_1-proteinase inhibitor.

Glycosaminoglycan	conc.	Lowest value of k_{ass} $(M^{-1} s^{-1})$	Maximal decrease in k_{ass} (n-fold)
none		1.3×10^7	-
HMW heparin[a]	2^d	4.5×10^4	290
LMW heparin[a]	54^d	3.2×10^5	40
dermatan sulfate[b]	e	2.6×10^6	5
chondroitin-4-sulfate[b]	e	2.6×10^6	5
chondroitin-6-sulfate[b]	e	2.6×10^6	5
hyaluromic acid[c]	f	6.5×10^6	2

[a] three sulfate groups per disaccharide unit (HMW, LMW = high or low molecular weight)
[b] one sulfate group per disaccharide unit
[c] no sulfate
[d] mol heparin / mol elastase
[e] 0.2 - 0.4 mg / ml reaction medium
[f] 1 mg / ml reaction medium

Table 1 shows that high molecular weight heparin is seven-fold more efficient in depressing k_{ass} and acts at a much lower concentration than low molecular weight heparin. The other glycosaminoglycans are less effective.

The heparin effect did not take place in the presence of 1 M NaCl or protamine sulfate. On the other hand, heparin did not depress k_{ass} for the reaction of α_1-proteinase inhibitor with porcine and bovine pancreatic trypsin or porcine pancreatic elastase.

Several heparins used in clinical care have been tested for their ability to affect the elastase-α_1-proteinase inhibitor binding (table 2). As can be seen, standard heparin, a mixture of high molecular weight chains, strongly impairs the inhibition of elastase by α_1-proteinase inhibitor when tested as a dose used in clinical care. Low molecular weight heparins are again less effective.

Table 2. Effect of heparin preparations used in clinical care on the elastase + α_1-proteinase inhibitor association. Three different standard heparins and two different low molecular weight heparins have been tested.

Heparin	units / ml reaction medium[a]	decrease in k_{ass} (n-fold)	delay time of inhibition (seconds)[d]
none			0.013
Standard heparins	0.5[b]	200-250	2.6 -3.2
Low molecular weight heparins	0.5[c]	14-15	~ 0.2

[a] these are close to the therapeutic concentration recommended by the pharmaceutical companies
[b] anticoagulant units
[c] antifactor Xa units
[d] given by $5 / k_{ass} [I_0]$ where $[I_0]$ is the plasma concentration of α_1-proteinase inhibitor ($\sim 30\ \mu M$)

DISCUSSION

The heparin effect evidenced here appears to be due to an electrostatic binding of the sulfated polysaccharide to the elastase molecule since (i) α_1-proteinase inhibitor does not react with this ligand [8], (ii) the effect is reversed by ionic strength and protamine sulfate (iii) proteinases such as trypsin and porcine pancreatic elastase which are much less basic than neutrophil elastase, do not undergo this effect, (iv) glycosaminoglycans which are less sulfated than heparin, are also less effective (table 1), (v) low molecular weight heparin which has a lower affinity for neutrophil elastase than high molecular weight heparin [8] is also less effective in depressing k_{ass}.

Glycosaminoglycans linked to proteins form the proteoglycans which are ubiquitous components of the extracellular matrix. The matrix of human lung contains, among others, dermatan sulfate, chondroitin sulfate and hyaluronic acid [9] which all depress k_{ass}. In the lung interstitum, the delay time of inhibition of neutrophil elastase by α_1-proteinase inhibitor might therefore be higher than 120 milliseconds, the time calculated from k_{ass} and the inhibitor concentration in the alveolar epithelial lining fluid [10].

Oxidation of α_1-proteinase inhibitor at the critical P_1 methionine residue of the active center considerably reduces its k_{ass} for neutrophil elastase : $k_{ass} = 7.6 \times 10^3$ $M^{-1} s^{-1}$ (ref.11). Due to its slow-binding inhibitor behavior, the oxidized protein is no longer able to prevent elastolysis *in vivo*. This oxidative impairment of inhibitory activity is thought to be one of the biochemical links between smoking and emphysema [1]. Here we show that intravenous administration of standard heparin likewise makes α_1-proteinase inhibitor to behave like a slow-binding inhibitor. With a 200-250-fold increase in the delay time of inhibition, the protein might no longer be fast-acting enough to prevent intravascular proteolysis of biologically important proteins such as coagulation factors, immunoglobulins, complement components and proteinase inhibitors which all have been shown to be cleaved *in vitro* by neutrophil elastase [1].

Release of neutrophil elastase occurs in the plasma of patients with septicemia[12, 13] as a result of neutrophil activation by endotoxins [14]. Activation of neutrophils also takes place during hemodialysis due to the contact of the cells with the dialyzer membrane[15]. Since heparin may be used as an anticoagulant drug in these two clinical situations, elastase mediated intravascular proteolysis may occur, as can be inferred from our data.It is worthwhile mentioning that decreased activities of a variety of coagulation factors have been observed in patients with severe septicemia [12, 13]. Moreover, Jordan and coworkers [16] recently reported that heparin considerably increases the rate of inactivation of antithrombin III by neutrophil elastase. Their finding, together with our data, suggests that if neutrophil activation takes place in patients under heparin therapy, massive degradation of antithrombin III may occur. It is noteworthy that Duswald et al. [13] found a 50 % antithrombin III inactivation in patients with severe septicemia. It was however not explicitely stated that the afore-mentioned patients with septicemia [12, 13] were under heparin therapy.

REFERENCES

1. J. G. Bieth, Elastase : catalytic and biological properties, in "Regulation of matrix accumulation", R. P. Mecham ed. Academic Press New York (1986).
2. J. C. Taylor and C. Mittman, "Pulmonary emphysema and proteolysis "Harcourt Brace Jovanovich, Publishers, Duarte (1987).
3. E. J. Campbell, Preventive therapy of emphysema : lessons from the elastase model, Am. Rev. Respir. Dis. 134 : 984 (1986).
4. K. Beatty, J. G. Bieth, and J. Travis, Kinetics of association of serine proteinases with native and oxidized α_1-proteinase inhibitor and with α_1-antichymotrypsin, J. Biol. Chem. 255 : 3931 (1980).
5. J. G. Bieth, Pathophysiological interpretation of kinetic constants of protease inhibitors, Bull. Europ. Physiopath. Respir. 16 (suppl.) : 183 (1980).

6. B. Casu, P. Oreste, G. Torri, G. Zoppetti, J. Choay, J. C. Lormeau, M. Petitou, P. Sinoy, The structure of heparin oligosaccharide fragments with high anti-(factor Xa) activity containing the minimal antithrombin III-binding sequence. Chemical and 13 C nuclear magnetic-resonance studies. <u>Biochem. J</u>. 197 : 599 (1981).

7. F Redini, J. M. Tixier, M Petitou, J. Choay, L. Robert, and W. Hornebeck, Inhibition of leucocyte elastase by heparin and its derivatives. <u>Biochem. J.</u> 252 : 515 (1988).

8. I. Danishefsky and R. Pixley, Effect of heparinon theinhibition of thrombin by alpha-1-proteinase inhibitor, Biochem. Biophys. Res. Commun. 91 : 862 (1979).

9. J.G. Clark, C. Kuhn, and R.P. Mecham. Lung Connective Tissue. <u>in </u>"International Review of Connective tissue research", D.A. Hall and D.S. Jackson eds. Academic Press. New York (1983).

10. F. Ogushi, R. C. Hubbard, G. A. Fells, M. A. Casolaro, D. T. Curiel, M. L. Brantly, and R. G. Crystal, Evaluation of the S-type of Alpha-1-antitrypsin as an *in vivo* and *in vitro* inhibitor of neutrophil elastase, <u>Am. Rev. Respir. Dis</u>. 137 : 364 (1988).

11. M. Padrines, M. Schneider-Pozzer, and J. G. Bieth, Inhibition of neutrophil elastase by α_1-proteinase inhibitor oxidized by activated neutrophils. <u>Am. Rev. Respir. Dis</u>. 139 : 783 (1989).

12. R. Egbring, W. Schmidt, G. Fuchs, K. Havemann, Demonstration of granulocytic proteases in plasma of patients with acute leukemia and septicemia with coagulation defects, <u>Blood</u>, 49 : 219 (1977).

13. K. H. Dustwald, M. Jochum, W. Schramm, H. Fritz, Released granulocytic elastase : an indicator of pathobiochemical alterations in septicemia after abdominal surgery, <u>Surgery</u>, 98 : 892 (1985).

14. A. O. Aasen, and K. Ohlsson, Release of granulocyte elastase in lethal canine endotoxin shock. <u>Hoppe-Seyler's Z. Physiol. Chem</u>. 359 : 683 (1978).

15. W. H. Hörl, H. B. Steinhauer, R. P. Schollmeyer, Plasma levels of granulocyte elastase during hemodialysis : effects of different dialyzer membranes, <u>Kidney Int.</u> 28 : 791 (1985).

16. R. E. Jordan, J. Kilpatrick, R. M. Nelson, Heparin promotes the inactivation of antithrombin by neutrophil elastase, <u>Science</u>, 237 : 777 (1987).

C5a RECEPTORS ON NEUTROPHILS AND MONOCYTES FROM CHRONIC DIALYSIS PATIENTS

Sharon L. Lewis

Department of Pathology
College of Nursing
University of New Mexico
School of Medicine
Albuquerque, NM 87131

INTRODUCTION

C5a is generated by the activation of the complement system with the cleavage of complement component C5 into C5a and C5b. C5a, an important _in vivo_ chemotactic factor and anaphylatoxin, was first identified as having an important role in dialysis patients by Craddock et al (1,2,3) in the 1970's. Previously various investigators had observed that leukopenia occurred during the first 30 minutes of hemodialysis (HD) with cellulose dialysis membranes (4,5,6,7). Craddock et al demonstrated that this leukopenia resulted from pulmonary sequestration of neutrophils (PMN) which was provoked by a complement component generated from contact of the patient's blood with the cellophane dialyzer (1,2,3). C5a became the likely suspect because it was known to induce neutrophil aggregation _in vivo_ and it was logically reasoned that PMN and monocytes would accumulate in the lungs, the first large capillary network encountered in circulation, following passage of blood through the dialyzer.

The purpose of this article is to provide a brief overview of the characteristics and functions of C5a. Then it will review research which has investigated C5a receptors and C5a-stimulated functional responses in PMN and monocytes from dialysis patients.

OVERVIEW OF C5a

C5a is a cationic glycoprotein containing 74 amino acids with a carboxyl terminal arginine. It is regulated by serum carboxypeptidase which cleaves the terminal arginine producing C5a des arg. Approximately 95% of human PMN and 70% of human monocytes have receptors for C5a (8). PMN have 150,000 - 200,000 C5a receptors per cell and

monocytes 80,000 to 100,000 C5a receptors per cell (9,10).
Binding of C5a to its receptor is specific, irreversible,
rapid (t 1/2 < 3 minutes), and saturable (9).

C5a plays a critical role in the acute inflammatory
response. As a anaphylatoxin it stimulates the secretion
of histamine and other components from mast cells and
basophils, the release of serotonin from platelets, and the
degranulation of eosinophils (11,12). As a chemotactic
factor, it mobilizes PMN and monocytes to the site of
infection and stimulates adherence, lysosomal enzyme
release, and oxidative metabolism in PMN and monocytes (2,
13-16). These responses occur after C5a binds to specific
high affinity receptors for C5a (9). C5a also stimulates
the production and release of interleukin-1 (IL-1) and
tumor necrosis factor (TNF) by macrophages (17,18).
Increased levels of IL-1 production have been demonstrated
in dialysis patients following the initiation of the HD
procedure (19).

C5a RECEPTOR MODULATION ON PMN AND MONOCYTES FROM DIALYSIS PATIENTS

Defective Chemotactic Response

A consistent finding in <u>in vitro</u> studies on dialysis
patients has been defective PMN chemotaxis regardless of
the chemoattractant used. (These studies are reviewed in
Lewis and Van Epps [20]). <u>In vivo</u> PMN chemotaxis, as
measured by the accumulation of cells in skin windows, has
also been shown to be impaired (21). Using blind-well
chemotactic chambers, we tested the <u>in vitro</u> chemotactic
response of dialysis patients' PMN to C5a (Figure 1). The
results indicated there was about a 50% decrease in the
chemotactic response from PMN of these patients.

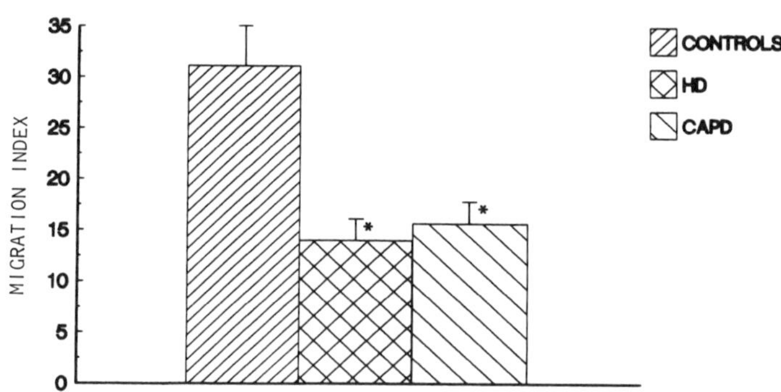

Figure 1. Comparison of the <u>in vitro</u> PMN chemotactic
response to C5a among controls (n=9), HD patients
(n=6), and CAPD patients (n=9). The results are
reported as the migration index expressed as the mean
± SEM. *p<0.01 or less as compared to control values.

The relationship between the defect in PMN chemotaxis
and increased susceptibility to infection in dialysis
patients remains speculative. Patients with defective PMN
chemotaxis are known to have an increased frequency of
bacterial infections (22). However, in dialysis patients
many other factors such as poor nutrition and dialysis
access sites are also major contributing factors to
infection. Nevertheless, the implication is that a
chemotactic defect could decrease the capacity of these
patients to defend against bacterial infections.

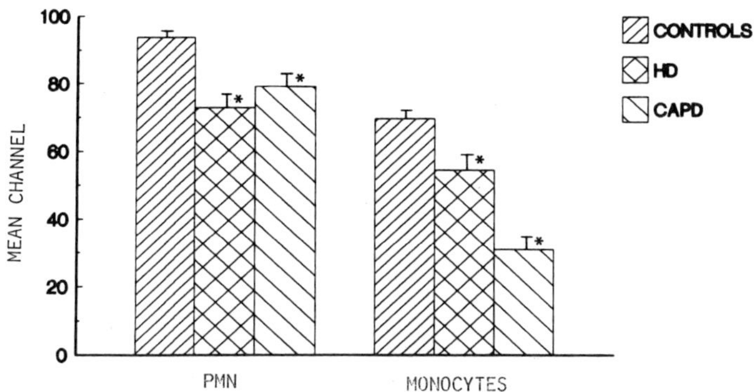

Figure 2. Binding of fluorescent C5a by PMN and
monocytes from controls (n=16), HD patients (n=13),
and CAPD patients (n=12). The results are reported as
mean (± SEM) percentage of positive cells. *p<0.005
or less as compared to control values.

C5a Receptor Modulation

A possible mechanism for decreased chemotactic
response is an alteration in chemotactic receptors. This
question was addressed in our studies where the binding of
several fluorescent chemotactic factors to PMN and
monocytes from dialysis patients was assessed using flow
cytometry (23). The results with C5a receptors indicated
there was a uniform marked reduction in the fluorescence
intensity of C5a-receptor-positive PMN from both HD and
continuous ambulatory peritoneal dialysis (CAPD) patients
as compared to controls (Figure 2). These shifts in
binding were also reflected in the mean fluorescence
intensity which paralleled shifts in the percentage of
positive cells. Compared to normal controls, the average
amount of C5a bound by PMN was decreased 25.0% for HD
patients and 16.0% for CAPD patients while the reduction in
monocytes was decreased 23.1% for HD patients and 55.6% for
CAPD patients.

<u>Determination of Complement Activation in Dialysis Patients</u>

Plasma C3a levels are a sensitive indicator of complement activation in whole blood. Comparison of plasma C3a levels in normal controls and dialysis patients indicated that both groups of dialysis patients had increased levels of C3a as compared to controls (Figure 3). Although all the HD patients enrolled in this study were dialyzed with cuprophane membranes, their plasma samples were obtained at least 44 hours after their last dialysis. Although plasma C3a levels were elevated in dialysis patients indicating chronic complement activation, no statistical correlation was found between plasma C3a levels and the percentage of C5a-positive PMN or monocytes in either patient group (23).

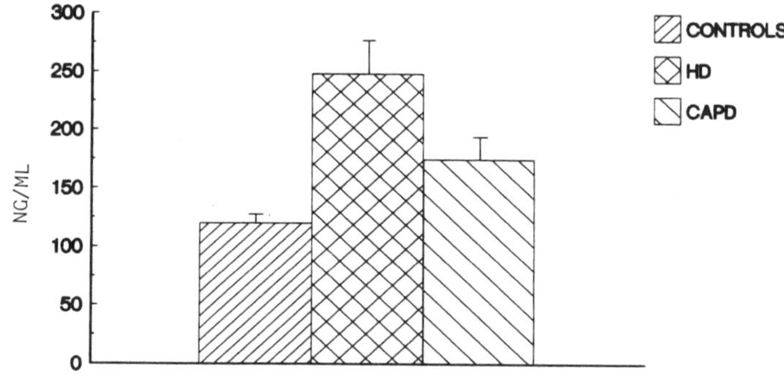

Figure 3. Comparison of plasma C3a levels in controls (n=8) with HD patients (n=7) (p<0.0545) and CAPD patients (n=17) (p<0.0585). Results are expressed as mean (+ SEM) ng/ml.

<u>Significance of C5a Receptor Modulation</u>

As stated earlier, during the HD procedure there is generation of C5a due to membrane-induced activation of the complement system. Chenoweth et al (24) found that the predialysis plasma C5a levels in HD patients were higher than in controls. Thus, these patients are exposed to high concentrations of C5a which may modulate the number of available C5a receptors on PMN and monocytes. Therefore, it is possible to explain the decreased responsiveness of PMN and monocytes from HD patients based on down-regulation of the C5a-receptor secondary to <u>in vivo</u> complement activation by the HD membrane. This explanation is supported by other studies involving deactivation of chemotactic-responsiveness cells (25,26,27,28,29). These studies showed that preexposure of PMN to chemotactic factor resulted in a decrease in the number of receptors available for subsequent chemotactic factor binding.

In CAPD patients it is more difficult to explain the decrease in C5a receptors because they are not exposed to HD membrane-induced complement activation. It may be that the indwelling peritoneal catheter and continual chemical stimulation of the dialysate solution has an activation effect on the complement system. This activation could produce enough C5a to ultimately block the C5a receptors on circulating cells. Another possible explanation for decreased C5a receptors in these patients is that all chronic renal failure patients, regardless of treatment, have decreased PMN and monocyte C5a-receptor expression.

These results show that there is C5a receptor modulation on PMN and monocytes from both CAPD and HD patients. Modulation of this chemotactic factor receptor may be an important contributing factor to the increased risk for infection by altering the responsiveness of PMN and monocytes to C5a generated during an inflammatory response.

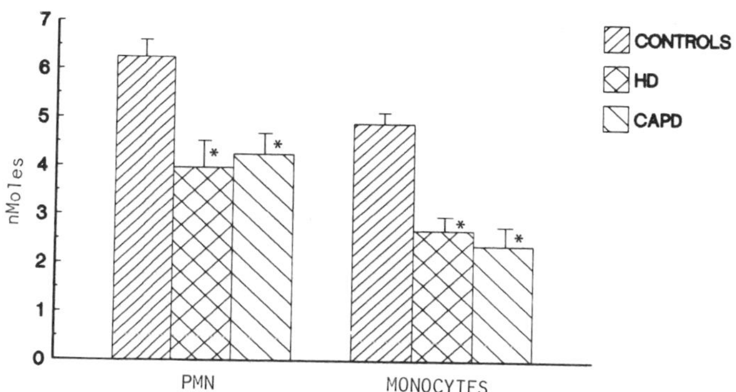

Figure 4. Release of superoxide anion from PMN and monocytes in response to C5a. Comparisons include controls (n=15), HD patients (n=15), and CAPD patients (n=16). Results are expressed as nMoles O_2^- produced/10^6 cells. *$p < 0.005$ or less as compared to control values.

CHEMOTACTIC FACTOR-INDUCED FUNCTIONAL RESPONSES

In addition to stimulating directional locomotion, chemotactic factors can also stimulate superoxide anion (O_2^-) generation and degranulation as well as prime PMN and monocytes so that they can produce enhanced quantities of toxic O_2 radicals (O_2^-, OH^-, H_2O_2) upon interaction with a second stimulus (30). These oxygen radicals are directly involved in killing target microorganisms. The microbicidal activity of H_2O_2 can be further augmented by myeloperoxidase (MPO) which is released from PMN granules. In the presence of chloride, MPO acts on H_2O_2 resulting in the production of highly toxic hypochlorous acid.

    To determine if C5a-mediated functional responses are
altered in dialysis patients, three parameters were
assessed including superoxide anion generation, H_2O_2
production, and MPO release (31). The ability of PMN and
monocytes from chronic dialysis patients to produce
superoxide in response to C5a was significantly lower than
that from normal controls (Figure 4). Compared to normal
controls, the PMN C5a-mediated response was decreased by
37% for HD patients and 32% for CAPD patients while the
monocyte C5a-stimulated response was decreased 45% for HD
patients and 51% for CAPD patients.

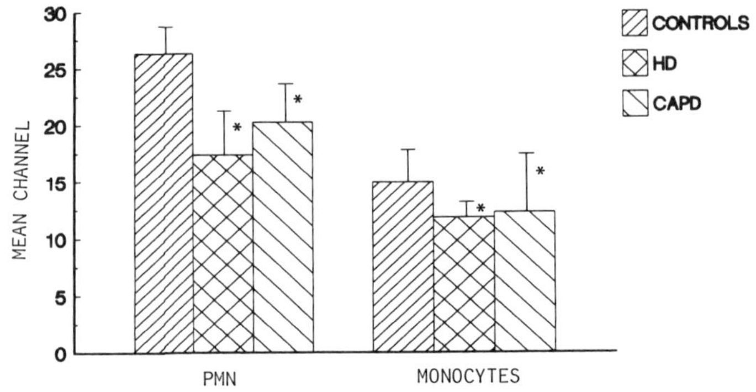

Figure 5. C5a-stimulated H_2O_2 production by PMN
and monocytes from normal controls (n=9), HD patients
(n=7), and CAPD patients (n=8). The results are
expressed as the mean (+SEM) channels comparing the
mean fluorescence channel for each sample to unstained
cells. *p<0.01 or less as compared to control values.

    To further evaluate this possible defect in oxidative
metabolism stimulated by chemotactic factors, H_2O_2
production by PMN and monocytes was assessed by determining
the oxidation of 2',7'-dichlorofluorescin in response to
C5a in a flow cytometric assay (31). As shown in Figure 5,
C5a-induced H_2O_2 production by PMN and monocytes from
dialysis patients was significantly less than that from the
normal controls.

    Degranulation was measured by evaluating MPO release
from PMN obtained from normal controls and dialysis
patients (Figure 6). The results indicate that
C5a-stimulated release of MPO from PMN from either group of
dialysis patients was about 80% of the control value.

These three sets of studies indicate that C5a-mediated superoxide generation, H_2O_2 production, and MPO release are altered. This would likely be the expected findings with C5a-stimulated responses since there is down-regulation of C5a receptors. Similar findings were demonstrated in a group of studies by Solomkin et al (32,33,34,35,36) on patients with trauma and infection where possible mechanisms for abnormal chemotactic and bactericidal ability of PMN were investigated. They found decreased _in vitro_ chemotactic responsiveness of PMN to C5a and decreased C5a binding to PMN. Plasma levels of C3a were elevated as compared to the controls and there was a linear relationship between plasma C3a levels and decreased _in vitro_ PMN chemotactic response to C5a. These studies indicate that deactivation and degranulation in response to high circulating levels of complement components may be involved in the abnormal PMN responses. A similar situation could be involved in dialysis patients where complement activation occurs.

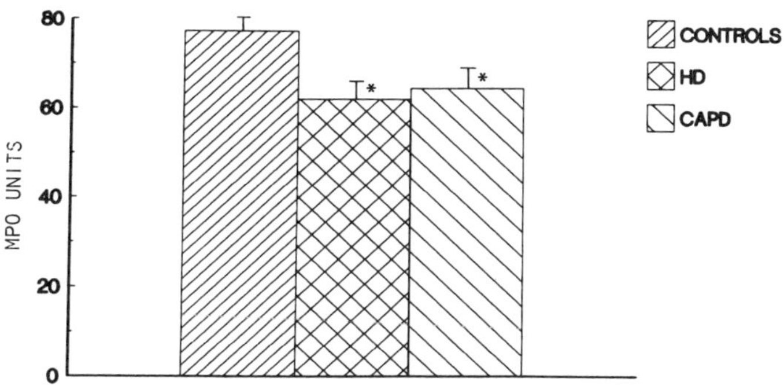

Figure 6. C5a-stimulated release of myeloperoxidase from PMN of controls (n=10), HD patients (n=11), and CAPD patients (n=8). The results are expressed as mean (+SEM) MPO units/10^6 cells. *p<0.02 or less as compared to control values.

The production of oxygen radicals and release of MPO are critical to the killing of microorganisms. The observed reduction of C5a-stimulated 0_2^- generation, H_2O_2 production, and MPO release from PMN and monocytes of dialysis patients may be a contributing factor in the increased susceptibility to infection in these patients.

Comparison of C5a Receptors of PMN and Monocytes from CAPD Patients

To determine the effect of peritoneal dialysis on C5a receptors, we obtained the peritoneal dialysate effluents (PDE) from the first exchange of CAPD patients, isolated the WBC, and compared the results of C5a binding to PMN and

monocytes/macrophages between the peripheral blood and
PDE. The results are shown in Figure 7. Interestingly,
there are significant differences with the PMN from PDE
demonstrating 63% less C5a binding than the peripheral
blood PMN while PDE macrophages bound 42% more C5a than
peripheral monocytes.

The explanation for these findings remains
speculative. It may be that the C5a receptors on PMN in
PDE are further down-regulated compared to the peripheral
blood PMN from these patients. Macrophages in the PDE
undergo further differentiation and maturation as compared
to peripheral blood monocytes. It may be that these
macrophages have more C5a receptors than peripheral
monocytes.

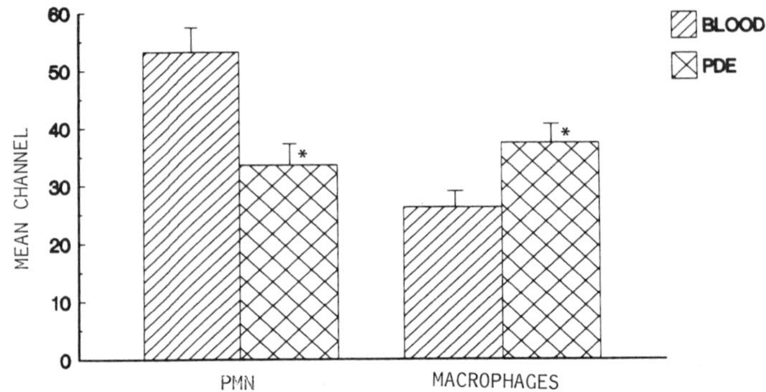

Figure 7. Comparisons of C5a receptors on PMN and
monocytes/macrophages from the blood and PDE of CAPD
patients (n=18). *p<0.001 by paired t-test.

C5a RECEPTOR MODULATION DURING HEMODIALYSIS

<u>Intradialysis Determination of Complement Activation</u>

Serial analysis of C3a and C5a done throughout the HD
procedure on patients using either cuprophane or
polyacrylonitrile dialyzers demonstrated that patients
using cuprophane dialyzers had significantly elevated
levels of C3a in the venous or efferent (dialyzer to
patient) line of the dialyzer after only 2 minutes of
dialysis (24). During HD, venous plasma C3a levels
continued to increase and became maximally elevated after
15 minutes. After this time, C3a concentrations gradually
declined but did not reach predialysis levels by the end of
dialysis. In contrast to the findings with C3a, plasma
levels of C5a did not increase strikingly until near the
end of HD procedure. These observations are compatible
with the fact that the C5a produced is rapidly metabolized
and cleared from circulation as well as binding to WBC and

174

other tissues, especially the liver (37,38). Detection of elevated levels of C5a during the later part of cuprophane dialysis suggested that the C5a-binding sites on both WBC and other tissues may eventually become saturated.

Reused cellulose-type dialyzers as well as polyacrylonitrile, polysulfone, and polycarbonate dialyzers have less ability to activate complement (24,39,40,41). The extent of complement activation paralleled the ability of these membranes to induce neutropenia.

The clinical sequelae of recurrent complement activation are not really known. There is an increase in intradialytic symptoms such as chest pain and back pain when new cuprophane dialyzers are used (42,43). There also appears to be indirect evidence that there is an increased morbidity in patients dialyzed with cuprophane membranes as compared to lower or non-complement activating membranes (42,44,45). Recurrent dialysis with complement-activating surfaces leads to greater complement activation (24,41), a lowered predialysis WBC count (41,46,47), a decrease in rebound leukocytosis (41), enhanced clearance of complement products (41), and possibly more intradialytic symptoms (43).

Analysis of C5a Receptors Throughout the HD Procedure

To study the specific effects of the dialysis membrane on C5a receptor modulation, we pumped peripheral blood from 5 different controls through various kinds of hemodialyzers and analyzed C5a receptors from WBC of blood samples obtained at 6 different time intervals (0,5,10,20,30, and 45 minutes). The results showed that the percentage of C5a-receptor-positive PMN and monocytes progressively decreased throughout the course of "dialysis" with an average of 34% for PMN and 48% for monocytes by 45 minutes (48).

To evaluate the binding of C5a to WBC from HD patients during the course of hemodialysis, blood samples were obtained at 4 intervals (0,0.5, 2, & 4 hours) from the afferent (arterial) line. These patients had all been on HD for more than 2 months, had used cuprophane dialyzers, and the studies were performed during the first use of the dialyzer. The results of these experiments indicated there was no difference in the percentage or mean channel of fluorescence of PMN or monocytes binding C5a at the 4 time intervals studied (48).

There are two possible explanations for the differences between the findings of C5a-receptor-positive cells from patients on HD and the in vitro dialysis system. Although there is evidence of enhanced complement activation during dialysis, significantly increased levels of plasma C5a in arterial blood samples have not been found during the course of HD (39,43,46). C5a is rapidly metabolized and cleared from circulation as well as binding to WBC and to other cells (37,46). Therefore, it is possible that C5a would be rapidly removed or inactivated in vivo and would not effectively block C5a receptors on

WBC to the same degree as observed in the _in vitro_ system
where there is no clearance mechanism and no other cells to
bind to other than the WBC in the whole blood sample.

A second possible explanation may relate to the
production of a serum chemotactic factor
inactivator/inhibitor (CFI) that could rapidly catabolize
or inactivate C5a (49,50). The presence of CFI could be
responsible for the failure of PMN and monocytes from HD
patients to show down-regulation of C5a receptors during
the course of dialysis, since CFI would serve to regulate
the level of active C5a. CFI would also serve to protect
chronically dialyzed patients from the adverse effects of
excessive C5a generation.

These findings suggest there are regulatory mechanisms
functioning in chronically hemodialyzed patients to protect
them from the adverse effects of excessive C5a generation
during HD. Although this may be an advantage during
dialysis, it may also be detrimental to host defense by
preventing the normal function of C5a in an inflammatory
response, thereby enhancing the patient's susceptibility to
infection.

POSSIBLE MECHANISMS INVOLVED IN DOWN-REGULATION OF C5a RECEPTORS

The mechanism of decreased C5a receptor expression in
dialysis patients is unknown. Preliminary studies done in
our laboratory have shown that pretreatment of normal PMN
with degranulating stimuli results in a decrease in C5a
binding. The mechanism of the "loss" of C5a in these
normal cells is unknown. A similar situation may exist in
chronic dialysis patients who are exposed to multiple
degranulating stimuli, at least during the hemodialysis
procedure. We have shown that when patient cells are
isolated and incubated in buffer or normal plasma, there is
no increase in C5a expression at any point analyzed from 30
minutes to 4 hours, implying that patient plasma factors
are not solely responsible for this phenomenon.

Fluorescence microscopy shows that receptors for C5a
are homogenously distributed on PMN. After C5a binding,
the ligand-receptor complex is internalized and mobilized
to the Golgi region (51). C5a is dissociated from the
receptor by lowering the pH and the receptor is then
returned to the plasma membrane while the ligand is
degraded. Reexpression of surface C5a receptors following
internalization by incubation with unlabelled C5a occurs
within 1 to 2 hours (51). We could find no difference in
the rate or percentage of C5a receptor reexpression between
controls and HD patients (Figure 8). HD patients return to
their "normal" levels of C5a receptor expression but do not
reach the level of receptor expression of normal control
cells.

Evaluation of C5a receptor internalization using flow cytometry and pH quenching of external fluorescence showed that receptors for both controls and HD patients were internalized by PMN in less than 15 minutes. There was no difference between controls and HD patients in the rate or percentage of receptors internalized.

Cyclic AMP (cAMP) is an important mediator of chemotactic factor stimulus-secretion coupling. It mediates lysosomal enzyme release and superoxide anion generation. Using a variety of stimuli (i.e. synthetic chemotactic factor formyl-met-leu-phe, prostaglandin E_2, phorbol myristate acetate, theophylline, and ionomycin), we could find no differences between HD patients and normal controls.

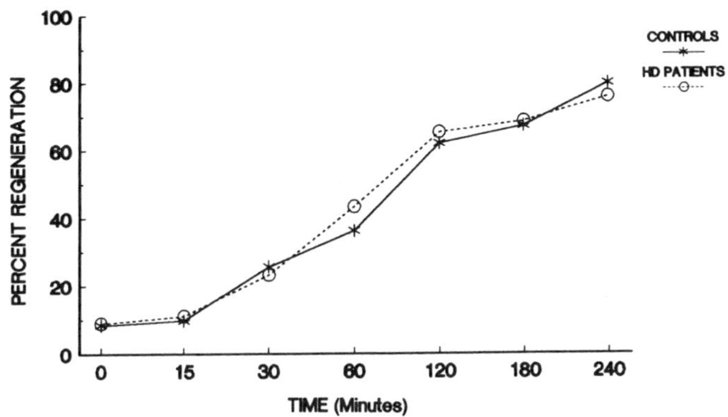

Figure 8. Comparison of C5a receptor reexpression between controls and HD patients. The results are expressed as the percentage of receptor reexpression relative to the baseline values.

In summary, these results demonstrate that there is decreased C5a receptor availability on PMN and monocytes from chronic dialysis patients as well as decreased responsiveness of these cells to C5a. These findings may contribute to these patients' increased risk to infection by reducing the ability of these cells to respond to C5a generated at a site of infection. These studies also indicate that C5a receptors on PMN and monocytes obtained from normal blood donors exposed to dialysis membranes are specifically blocked or internalized. In contrast, even though PMN and monocytes from HD patients are continuously exposed to C5a generated during the dialysis procedure, they do not show a decrease in the binding of C5a. These findings suggest the existence of adaptive mechanisms which may function in chronically dialyzed patients to prevent some of the adverse effects of excessive C5a generation during HD.

This work was supported in part by the National Kidney Foundation and DHHS grant number NR00001 from the National Center for Nursing Research.

References

1. P.R. Craddock, J. Fehr, K.L. Brigham, and R.S. Kronenberg, Complement and leukocyte-mediated pulmonary dysfunction in hemodialysis, New Engl J Med 296:769-774 (1977).
2. P.R. Craddock, D. Hammerschmidt, J.G. White, A.P. Dalmasso, and H.S. Jacob, Complement (C5a)-induced granulocyte aggregation in vitro, J Clin Invest 60:260-264 (1977).
3. P.R. Craddock, J. Fehr, A.P. Dalmasso, K.L. Brigham, and H.S. Jacob, Hemodialysis leukopenia: Pulmonary vascular leukostasis resulting from complement activation by dialyzer cellophane membranes, J Clin Invest 59:879-888 (1977).
4. L.S. Kaplow and J.A. Goffinet, Profound neutropenia during the early phase of hemodialysis, J Am Med Assoc 203:133-135 (1968).
5. M. Toren, J.A. Goffinet, and L.S. Kaplow, Pulmonary bed sequestration of neutrophils during hemodialysis, Blood 36:337-340 (1970).
6. T. Gral, P. Schroth, J.R. DePalma, and A. Gordon, Leukocyte dynamics with three types of hemodialyzers, Trans Amer Soc Artif Int Organs 15:45-49 (1969).
7. L.H. Brubaker and K.D. Nolph, Mechanisms of recovery from neutropenia induced by hemodialysis, Blood 38:623-631 (1971).
8. D.E. Van Epps and D.E. Chenoweth, Analysis of the binding of fluorescent C5a and C3a to human peripheral blood leukocytes, J Immunol 132:2862-2867 (1984).
9. D.E. Chenoweth and T.E. Hugli, Demonstration of a specific C5a receptor on intact human polymorphonuclear leukocytes, Proc Natl Acad Sci (USA) 75:3943-3947 (1978).
10. D.E. Chenoweth, M.G. Goodman, and W.O. Weigle, Demonstration of a specific receptor for human C5a anaphylatoxin on murine macrophages, J Exp Med 156:68-78 (1982).
11. T.E. Hugli and H.J. Muller-Eberhard, Anaphylatoxins: C3a and C5a, Adv Immunol 26:1-53 (1978).
12. T.E. Hugli, The structural basis for anaphylatoxin and chemotactic functions of C3a, C4a, and C5a, CRC Critical Reviews in Immunology 1:321-366 (1981).
13. D.E. Chenoweth and T.E. Hugli, Human C5a and C5a analogs as probes of the neutrophil C5a receptor, Mol Immunol 17:151-161 (1980).
14. H.N. Fernandez, P.M. Henson, A. Otani, and T.E. Hugli, Chemotactic response to human C3a and C5a anaphylatoxins. I. Evaluations of C3a and C5a leukotaxis in vitro and under simulated in vivo conditions, J Immunol 120:109-115 (1978).

15. C. Gerard, D.E. Chenoweth, and T.E. Hugli, Response of human neutrophils to C5a: a role for the oligosaccharide moiety of human C5a des arg-74 but not of C5a in biologic activity, $\underline{J}$ $\underline{Immunol}$ 127:1978-1982 (1981).

16. R.O. Webster, S.R. Hong, R. Johnston, and P. Henson, Biological effects of human complement fragments C5a and C5a$_{des\ arg}$ on neutrophil function, $\underline{Immunopharmacology}$ 2:201-219 (1980).

17. M.G. Goodwin, D.E. Chenoweth, and W.O. Weigle, Induction of interleukin I secretion and enhancement of humoral immunity by binding of human C5a to macrophages surface C5a receptors, $\underline{J}$ $\underline{Exp}$ $\underline{Med}$, 156:912-917 (1982).

18. S. Okusawa, K.B. Yancey, J.W. van der Meer, S. Endres, G. Lonnemann, K. Hefter, M.M. Frank, J.F. Burke, C.A. Dinarello, J.A. Gelfand, C5a stimulates secretion of tumor necrosis factor from human mononuclear cells $\underline{in}$ $\underline{vitro}$. Comparison with secretion of interleukin 1 beta and interleukin 1 alpha, $\underline{J}$ $\underline{Exp}$ $\underline{Med}$ 168:443-448 (1988).

19. M. Blumenstein, B. Schmidt, R.A. Ward, H.W.L. Ziegler-Heitbrock, and H.J. Gurland, Altered Interleukin-1 Production in Patients Undergoing Hemodialysis, $\underline{Nephron}$ 50:277-281 (1988).

20. S.L. Lewis and D.E. Van Epps, Neutrophil and monocyte function in chronic dialysis patients (In-depth review), $\underline{Am}$ $\underline{J}$ $\underline{Kidney}$ $\underline{Dis}$ 9:381-395 (1987).

21. Z. Hanicki, T. Cichocki, Z. Komorowska, W. Sulowica, and O. Smolenski, Some aspects of cellular immunity in untreated and maintenance hemodialysis patients, $\underline{Nephron}$ 23:273-275 (1979).

22. M.E. Miller, Pathology of chemotaxis and random migration, $\underline{Sem}$ $\underline{Hematol}$ 12:59-81 (1975).

23. S.L. Lewis, D.E. Van Epps, and D.E. Chenoweth, C5a receptor modulation on neutrophils and monocytes from chronic hemodialysis and peritoneal dialysis patients, $\underline{Clin}$ $\underline{Nephrol}$ 26:37-44 (1986).

24. D.E. Chenoweth, A.K. Cheung, and L.W. Henderson, Anaphylatoxin formation during hemodialysis: Effects of different dialyzer membranes, $\underline{Kidney}$ $\underline{Int}$ 24:764-769 (1983).

25. R.D. Nelson, R.T. McCormack, J.D. Fiegel, and R.L. Simmons, Chemotactic deactivation of human neutrophils: Evidence for nonspecific and specific components, $\underline{Infect}$ $\underline{Immun}$ 22:441-444 (1978).

26. R.D. Nelson, V.D. Fiegel, M.J. Herron, and R.L. Simmons, Chemotactic deactivation of human neutrophils: Relationship to loss of cytotoxin receptor function and temporal nature of phenomenon, $\underline{J}$ $\underline{Retic}$ $\underline{Soc}$ 28:285-294 (1980).

27. R.D. Nelson, J.M. Gracyk, V.D. Fiegel, M.J. Herron, and D.E. Chenoweth, Chemotactic deactivation of human neutrophils: Protective influence of phenylbutazone, $\underline{Blood}$ 58:752-758 (1981).

28. J.T. O'Flaherty, D.L. Kreutzer, H.J. Showell, G. Vitkauskas, E.L. Becker, and P.A. Ward, Selective neutrophil desensitization to chemotactic factors, $\underline{J}$ $\underline{Cell}$ $\underline{Biol}$ 80:564-572 (1979).

29. S.J. Sullivan and S.H. Zigmond, Chemotactic peptide
     receptor modulation in polymorphonuclear
     leukocytes, <u>J Cell Biol</u> 85:703-711 (1980).
30. J. Bender, L. McPhail, and D.E. Van Epps, Exposure of
     human neutrophils to chemotactic factors
     potentiates activation of the respiratory burst
     enzyme, <u>J Immunol</u> 130:2316-2323 (1983).
31. S.L. Lewis, D.E. Van Epps, and D.E. Chenoweth,
     Alterations in chemotactic factor-induced
     responses of neutrophils and monocytes from
     chronic dialysis patients, <u>Clinical Nephrology</u>
     30:63-72 (1988).
32. J.S. Solomkin, M.P. Bauman, R.D. Nelson, and R.L.
     Simmons, Neutrophil dysfunction during the course
     of intra-abdominal infection, <u>Ann Surg</u> 194:9-17
     (1981).
33. J.S. Solomkin, M.K. Jenkins, R.D. Nelson, D.E.
     Chenoweth, and R.L. Simmons, Neutrophil
     dysfunction in sepsis. II. Evidence for the
     role of complement activation products in
     cellular deactivation, <u>Surg</u> 90:319-327 (1981).
34. J.S. Solomkin, L.A. Cotta, J.K. Brodt, J.W. Hurst, and
     C.K. Ogle, Neutrophil dysfunction in sepsis.
     III. Degranulation as a mechanism for
     nonspecific deactivation, <u>J Surg Res</u> 36:407-412
     (1984).
35. J.S. Solomkin, L.A. Cotta, J.D. Ogle, J.K. Brodt, C.K.
     Ogle, P.S. Satoh, J.M. Hurst, and J.W. Alexander,
     Complement-induced expression of cryptic
     receptors on the neutrophil surface: A mechanism
     for regulation of acute inflammation in trauma,
     <u>Surg</u> 96:336-344 (1984).
36. R.M. Antrum and J.S. Solomkin, Monocyte dysfunction in
     severe trauma: Evidence for the role of C5a in
     deactivation, <u>Surgery</u> 100:29-37 (1986).
37. R.O. Webster, G.L. Larsen, and P.M. Henderson, <u>In vivo</u>
     clearance and tissue distribution of C5a and <u>C5a</u>
     desarginine complement fragments in rabbits, <u>J
     Clin Invest</u> 70:1177-1183 (1982).
38. D.J. Weisdorf, D.E. Hammerschmidt, H.S. Jacob, and
     P.R. Craddock , Rapid <u>in vivo</u> clearance of C5a
     des arg: a possible protective mechanism against
     complement-mediated tissue injury, <u>J Lab Clin Med</u>
     98:823-830 (1981).
39. D.E. Chenoweth, A.K. Cheung, and L.W. Henderson,
     Anaphylatoxin formation during hemodialysis:
     Comparison of new and re-used dialyzers, <u>Kidney
     Int</u> 24:770-774 (1983).
40. D.F. Stroncek, P. Keshaviah, P.R. Craddock, and D.E.
     Hammerschmidt, Effect of dialyzer reuse on
     complement activation and neutropenia in
     hemodialysis, <u>J Lab Clin Med</u> 104:304-311 (1984).
41. R.M. Hakim, D.T. Fearon, and J.M. Lazarus,
     Biocompatibility of dialysis membranes: Effects
     of chronic complement activation, <u>Kidney Int</u>
     26:194-200 (1984).
42. D.V. Bok, L. Pascual, C. Herberger, R. Dawyer, and
     N.W. Levin, Effects of multiple use of dialyzers
     on intradialytic symptoms, <u>Proc Clin Dial Trans
     Forum</u> 10:92-98 (1980).

43. P. Ivanovich, D.E. Chenoweth, R. Schmidt, H. Klinkman, L.A. Boxer, H.S. Jacob, and D.E. Hammerschmidt, Symptoms and activation of granulocytes and complement with 2 dialysis membranes, Kidney Int 24:758-763 (1983).

44. K.S. Kant, V.E. Pollak, M. Cathey, D. Goetz, and R. Berlin, Multiple use of dialyzers: Safety and efficacy, Kidney Int 19:728-738 (1981).

45. A.J. Wing, F.P. Brunner, H.O.A. Brynger, C. Chantler, R.A. Donckerwolcke, H.J. Gurland, C. Jacobs, and N.H. Selwood, Mortality and morbidity of reusing dialyzers, Br Med Journal (abstract) 23:853 (1978).

46. R.M. Hakim and E.G. Lowrie, Hemodialysis associated neutropenia and hypoxemia: The effect of dialyzer membrane materials, Nephron 32:32-39 (1982).

47. N.A. Hoenich, D.N.S. Kerr, M.K. Ward, P. Aljama, and M. Sussman, Two special properties of polyacrylonitrile membrane suitable for reuse and biocompatibility, Cont Dial 5:31-41 (1984).

48. S.L. Lewis, D.E. Van Epps, and D.E. Chenoweth, Leukocyte C5a receptor modulation during hemodialysis and following exposure to dialysis membranes, Kidney Int 31:112-120 (1987).

49. J.R. McCormick, D.L. Kreutzer, H.J. Keating, J. Hupp, A. Despins, and M. Moore, Alterations in activities of anaphylatoxin inactivator and chemotactic factor inactivator during hemodialysis, Am J Pathol 109:282-287 (1982).

50. S.E. Goldblum, D.E. Van Epps, and W.P. Reed, Serum inhibition of C5 fragment-mediated polymorphonuclear leukocyte chemotaxis associated with chronic hemodialysis, J Clin Invest 64:255-264 (1979).

51. D.E. Van Epps, S. Simpson, J.G. Bender and D.E. Chenoweth, Regulation of C5a and formyl peptide receptor expression on human PMN, J Immunol (in press).

RESPECTIVE INFLUENCE OF UREMIA AND HEMODIALYSIS ON WHOLE BLOOD
PHAGOCYTE OXIDATIVE METABOLISM, AND CIRCULATING INTERLEUKIN-1 AND TUMOR
NECROSIS FACTOR

Béatrice Descamps-Latscha, André Herbelin, Anh Thu Nguyen and P. Urena*

INSERM U25 and CNRS UA 122, *INSERM U90, Hôpital Necker, Paris

INTRODUCTION

Evidence has long accumulated that uremia is associated with defects in immune
response capacity. Abnormalities in both T cell mediated immune responses and
granulocyte production, kinetics and function have been reported in patients with end stage
renal failure who were not yet dialysed or were undergoing long-term hemodialysis.
Several hypotheses have been put forward to explain this state of immunodeficiency
associated with chronic uremia and the changes in humoral and cellular immunity related to
the hemodialysis procedure (recently reviewed in 1).

In the past years our laboratory has been interested in the study of the influence of
hemodialysis on phagocyte oxidative metabolism, focusing on the role of dialysis
membrane bioincompatibility and on the possible definition by this phagocytic cell function
of biocompatible materials (2-4). More recently our investigation has been extended to the
possible role of interleukin-1 (IL-1) and tumor necrosis factor (TNF) (5), cytokines mainly
derived from monocytes and macrophages, which participate in the development of
immune response and exert a broad range of activities associated with both acute and
chronic inflammatory reactions.

The present report will review our recent studies aimed at determining whether the
uremic state per se or changes induced by the hemodialysis procedure could be responsible
for modulating phagocyte oxidative metabolism, and IL-1 and TNF productions. In an
attempt to closely reflect the *in vivo* situation we conducted this investigation directly
within whole blood for phagocyte oxidative metabolism investigation and in the plasma for
cytokine determination.

Phagocyte oxidative metabolism

Molecular basis of phagocyte derived reactive oxygen intermediate (ROI) production

It is now well documented that following appropriate membrane stimulation by
particulate or soluble agents, both polymorphonuclear (PMN) and mononuclear (MN)
phagocytes undergo a so-called respiratory burst (6) which leads to the production of
reduced oxygen intermediates (ROI) derived from the superoxide anion (O_2^-), i.e., the
hydrogen peroxide (H_2O_2) the hydroxyl radical (OH.) and the singlet oxygen ($^1O_2$).
Altogether these highly reactive molecules participate in host resistance against pathogens
(6) and tumoral cells (7) but may also be involved in host reactions towards normal
structures (8).

At the molecular level, stimulation of the respiratory burst involves activation of an
electron transport , the NADPH oxidase chain, which is dormant in the resting phagocyte

but, following activation, becomes capable of reducing molecular oxygen to $O2^{\cdot}$ in the presence of NADPH (9). Distinct mechanisms can be involved in the activation of the oxidase complex depending on the nature of the stimulus and its interaction pathway with the phagocyte membrane (10). Interestingly, the phagocyte respiratory burst can be triggered not only by bacteria, foreign particles and chemical substances but also by more physiological compounds, such as antibodies (11,12), immune complexes and, which is particularly relevant in the present context of hemodialysis, complement activated components (13). These findings outline the importance of using distinct stimulating agents when investigating phagocyte oxidative response potential.

Since the original description by Allen in 1972 (14) that phagocyte respiratory burst yields excited products that relax by photon emission or chemiluminescence (CL) and the subsequent demonstration that this native CL emission can easily be amplified by introducing high quantum yield substrates in the reaction, such as the cyclic hydrazide, luminol and the acidium salt, lucigenin, a consensus has now been reached that CL determination represents the most sensitive and adequate test for evaluating phagocyte oxygenation activities (15).

Likewise, it has now been well documented that investigation of CL directly within whole blood represents an ultrasensitive means for evaluating the overall oxidative response potential of PMN and MN phagocytes (16). A recent overview of the results obtained by our group in the application of this CL methodology for evaluating the role of phagocyte oxidative metabolism in allograft immunity and immunopathology is presented in ref. 8.

Phagocyte oxidative metabolism in long-term hemodialysis uremic patients

In 1985, we first demonstrated that activation of phagocyte oxidative metabolism can be observed in whole blood of hemodialysed patients (2). Briefly, 35 uremic patients undergoing long-term hemodialysis were studied, including 22 dialysed with cellulosic cuprophan (CUP) and 13 with synthetic polyacrilonitrile (PAN) membrane equipped dialysers. The phagocyte CL production was similarly monitored in microamounts of whole blood taken at various times during the dialysis session. Results demonstrated that in CUP dialysed patients, the CL production considerably increases in the early phase of the dialysis session (15 min after the start) but returns to its predialysis level by the end of the session. By contrast whole blood CL production remains unchanged in patients dialysed with the PAN membrane. This improved biocompatibility of PAN over CUP membranes was further demonstrated in our *in vitro* study on the effect of these membranes on CL production of whole blood from normal subjects. In this *in vitro* model we indeed showed that the kinetics of phagocyte oxidative metabolism activation induced by CUP membranes closely follows that observed *in vivo* and that there is an almost linear relationship between the CL level and the CUP membrane surface in contact with the blood.

Study of the respective participation of plasma factors and phagocytic cells in the whole blood CL reaction showed that both in the *in vivo* and in the *in vitro* models, PMN and MN phagocytes demonstrate normal oxidative responses when tested in the absence of plasma. In the presence of autologous plasma obtained at the 15th min of the session in hemodialysed patients or from normal blood incubated *in vitro* with dialysis membranes, a significant activation of ROI production is observed with CUP but not PAN membranes provided plasma complement had not been previously heat inactivated (17).

Further investigation of the possible relationships between leukopenia, activation of complement and phagocyte oxidative metabolism was conducted in a multicentric trial comparing two groups of long-term hemodialysis patients, treated with CUP or PAN membrane equipped dialysers and clinically well matched (4). No linear correlation could be demonstrated between the C3a increase and the neutropenia level observed in the CUP membrane dialysed group. By contrast a such correlation could easily be demonstrated between the amount of C3a generated following dialysis with CUP membranes and the subsequent increase of basal whole blood CL or decreased CL responses to PMA and fMLP. From these findings we concluded that the stimulation of whole blood ROI production observed during the initial period of the dialysis session in CUP but not PAN dialysed patients may be linked to the presence of a plasma factor derived from complement activation.

Other studies in the literature have shown that the passage of blood through hemodialysis circuits may diversely influence phagocyte oxidative metabolism, depending upon the time during the dialysis session (18,19) and that these changes are variably influenced by the patient's serum (18,20) or other factors such as the degree of hyperparathyroidism (21). Interestingly, in a recent study Markert et al. (22) evaluated the biocompatibility of dialysis membranes by both luminol and lucigenin amplified CL assays. Within a larger panel of dialysis membranes they also observed the improved bio-compatibility of PAN over CUP membranes and demonstrated the requirement of plasma in CUP induced PMN CL activation. Moreover, they further demonstrated that this dialysis membrane dependent PMN oxidative activation is in fact associated with impairement of other phagocyte functions such as aggregation, chemotaxis and adherence.

Phagocyte oxidative metabolism in first-dialysis uremic patients

The possibility that uremia per se may be responsible for a state of immunodeficiency has been mainly investigated in T cell-mediated functions (23). In respect to phagocytic cells several defects in chemotactic, adherence, and bactericidal activities have been reported and related to the high susceptibility of uremic patients to bacterial infections.

In order to better determine the respective role of uremia and of the hemodialysis procedure on the above described changes in phagocyte oxidative response potential we recently studied whole blood CL production in not yet dialysed uremic patients undergoing their first dialysis session. Fifteen uremic patients first dialysed with CUP membranes were tested before (t0) and at the 15th min. (t15) of the dialysis session. Results were compared to those previously reported in long-term hemodialysis patients (2). As shown on figure 1, first-dialysis uremic patients also demonstrate significant leukopenia in the early phase of the dialysis session, although its degree is much less pronounced than in long-term hemodialysis patients (p<0.001).

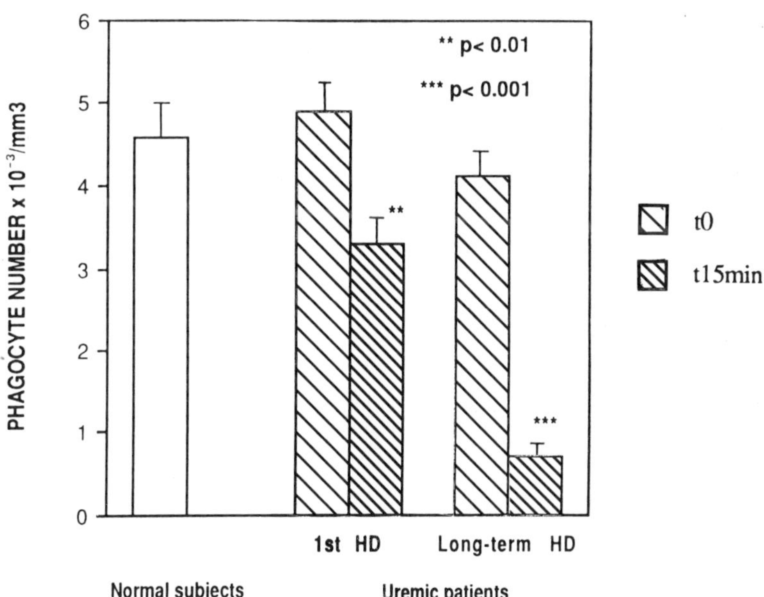

Figure 1. Influence of a single dialysis session on circulating phagocytes (mean number ± SEM) in 15 first and 22 long-term hemodialysis (HD) patients treated with CUP membrane equipped dialysers

In coincidence with the nadir of leukopenia, a moderate rise in whole blood resting CL production is also found in these patients (Figure 2); however, here again the CL increase between t0 and t15 is significantly lower than in long-term hemodialysis patients (p<0.001).

Study of whole blood stimulated CL responses to a panel of agents (figure 3) shows that compared to normal subjects, first-dialysis patients have significantly decreased CL response capacities to particulate agents (latex and opsonised zymosan) and among soluble stimuli to PMA but not fMLP.

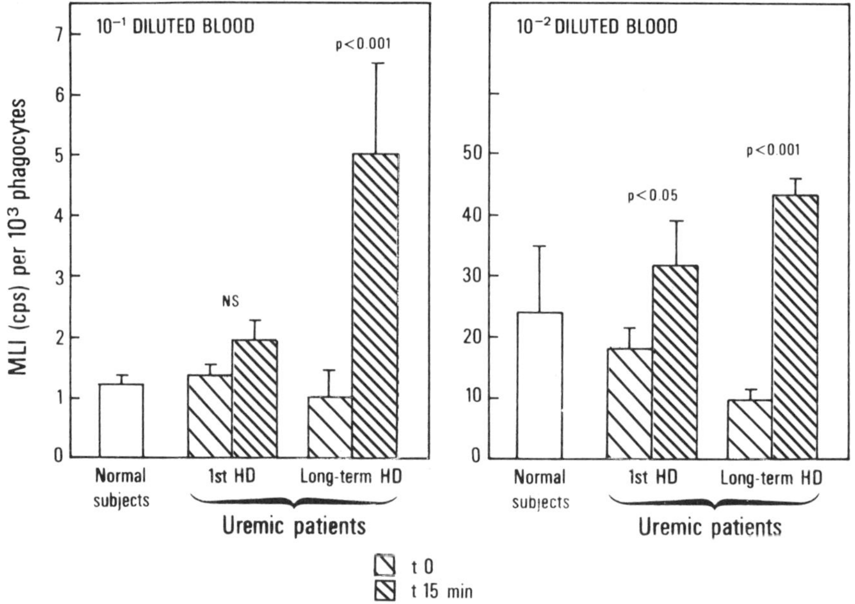

Figure 2. Influence of a single dialysis session on resting CL production (mean maximal light intensity (MLI) ±SEM) in 10⁻¹ and 10⁻² diluted blood samples taken before (t0) and 15 min after the start of dialysis session in 15 first- and 22 long-term hemodialysis (HD) uremic patients treated with CUP equipped dialyzers.

Lastly, as shown on Figure 4 the further decreasing effect of a single dialysis session on whole blood CL response capacity is again much more marked in long-term than in first-dialysis uremic patients.

Up to our knowledge this is the first reported study comparing the effects of first and long-term dialysis on this phagocyte oxidative function. Although the marked differences observed between the two groups of patients still remain to be further elucidated, it is interesting to suggest that a single session is already capable to induce minor phagocyte alterations and that these will be further amplified following the chronic dialysis treatment.

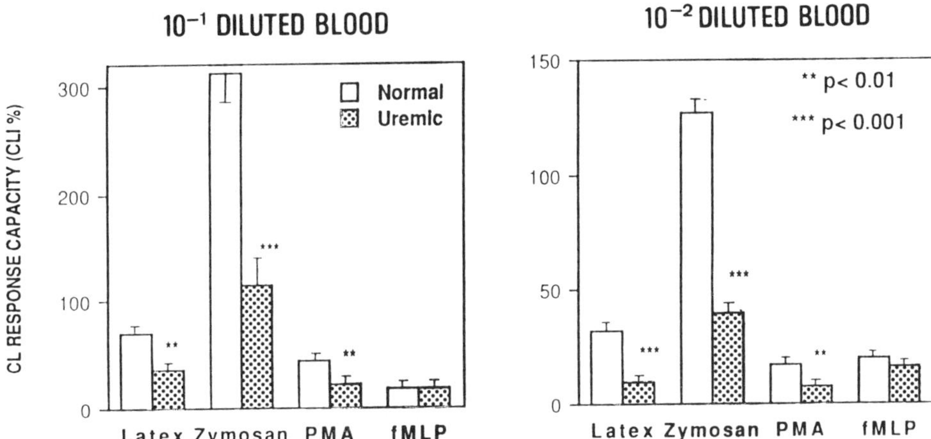

Figure 3. Influence of uremia on whole blood CL response capacity (mean CLI ± SEM) to a panel of particulate (latex, opsonised zymosan) and soluble (PMA, fMLP) stimulating agents. Results obtained in 15 not yet dialysed uremic patients are compared to those in 84 normal subjects*

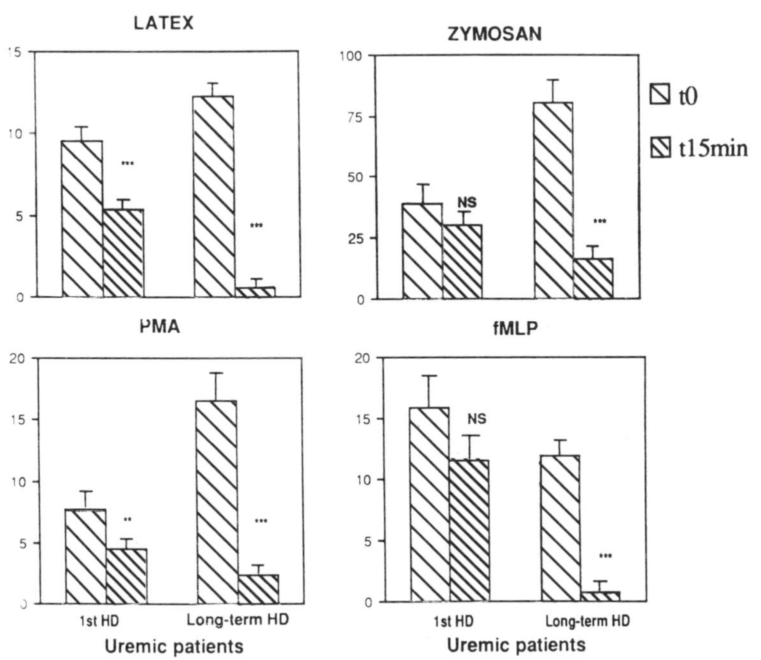

Figure 4. Influence of a single dialysis session on whole blood CL response capacity (mean CLI ± SEM) in 15 first and 22 long term hemodialysis (HD) patients.*

$$*CLI = 100 \times \frac{\text{stimulated CL - basal Cl}}{\text{basal CL}}$$

Whatever its mechanisms it is now evident that this increased ROI production triggered by bioincompatible dialysis membranes is indeed the reflection of an acute phase inflammatory reaction and may largely participate to the pathological changes and clinical symptoms observed in hemodialysed patients.

Among the biological effects of ROI, one may successively recall: the capacity of $O_2^{\cdot-}$ to induce a chemotactic factor from a plasma precursor (24), since this factor could contribute to PMN sequestration in the lung; the selective inactivating effect of ROI on the elastase inhibitor alpha-1-antitrypsin (25) since it could explain its remarkable increase, alone or complexed with elastase during dialysis (26); lastly the potential of ROI to activate the latent PMN collagenase (27) is suggestive of their important contribution in tissue damage.

Among the clinical changes which could be ascribed to ROI one may recall their predilective deleterious effect on cartilage structures (28) since it could participate in the articular damage often seen in long-term hemodialysed patients. Similarly the exquisite sensitivity of DNA to ROI which is often evoked in the possible role of these highly reactive molecules in aging could also account for the skin lesions (29) and the most early aging process frequently observed in these patients and sometimes strikingly referred to as the "shrinking man" syndrome (30).

Interleukin-1 and tumor necrosis factor

It is now well documented that Interleukin 1 (IL-1) and tumor necrosis factor (TNF) which are mainly produced by monocytes participate at several steps of the immune response and mediate many of the acute phase changes of the inflammatory reaction.

The potential effects of hemodialysis on IL-1 production were first outlined by Henderson et al. (31) in 1983. In a so-called IL-1 hypothesis, these authors attributed to IL-1 most dialysis-related events such as the rise in body temperature occurring during and after the dialysis session, the sleep tendency, the increased catabolism and protein breakdown, the fibroblast proliferation induced collagen deposition, the frequent headache which could be due to increased PGE2 production in cerebral tissue, the hypotension which could be explained by increased PGI2 production by endothelial cells and, lastly, the metabolic changes including hypozincemia and hypercupremia known to be induced by IL-1 (32).

Among the mechanisms responsible for this dialysis induced production of IL-1, dialysis membrane related factors such as monocyte adherence and complement activation (32,33) and dialysate associated compounds such as endotoxin contamination (32), and acetate (34) have focused the attention of some investigators.

In 1986 we first reported a preliminary study showing that plasma IL-1 is significantly increased in long term hemodialysis patients and that this was observed regardless of the nature of the dialysis membrane (17). Other groups measuring the capacity of monocytes to release IL-1 also reported, an increased IL-1 activity in unstimulated monocytes isolated from long term hemodialysis patients (35). Interestingly enough this was observed for the excreted form of IL-1 as well as for the intracellular and membrane associated form of IL-1 (36).

By contrast and although TNF shares with IL-1 remarkable similarities, both in its mode of induction and in its biological properties (37,38), its role in hemodialysis induced changes has not yet been extensively investigated.

In a recent study we have attempted to determine the possible relationships between circulating IL-1 and TNF in hemodialysis-related events and the respective influence of uremia and hemodialysis on the production of these two major monocyte-derived cytokines (5). Briefly, both IL-1 and TNF were measured by their biological activities i.e., respectively for IL-1 by its capacity to costimulate PHA induced thymocyte proliferation in the lymphocyte activating factor assay and for TNF by its cytolytic activity towards the L929 tumor cell line, as well as by their immunoreactivity as measured by a commercial ELISA for IL-1β and by a soluble RIA developed by Prof. P. Franchimont and IRE-Medgenix (Liège, Belgium), for TNFα.

Results shown on figures 5 and 6 can be summarized as follows:
1° both IL-1 and TNF are significantly increased in the plasma of long-term hemodialysed patients as compared to normal subjects, whereas in first-dialysis uremic patients, TNF but not IL-1, can be detected in the plasma (Figure 5).

2° During a single dialysis session, a significant increase in IL-1 is observed at the end of the session whereas TNF remains at its predialysis level (Figure 6).

Lastly, in both first and long-term dialysis patients no significant influence of the nature of the dialysis membrane could be demonstrated (results not shown), thus suggesting that the observed changes in the cytokine levels are not exclusively secondary to the activation of complement.

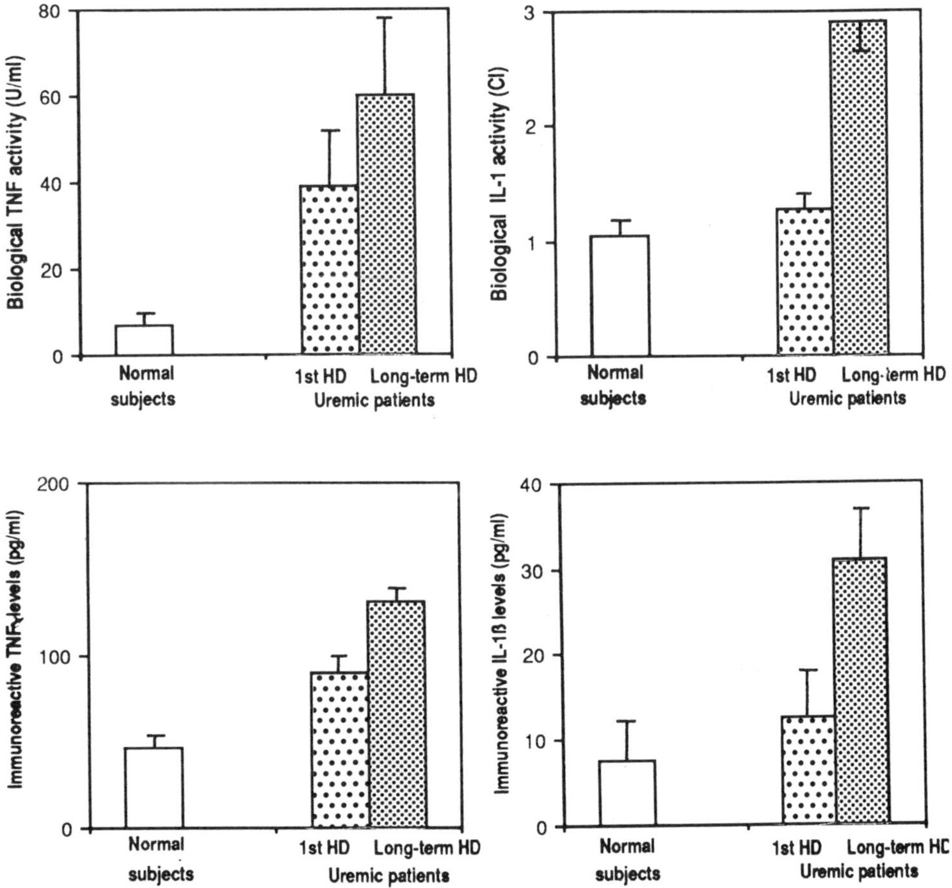

Figure 5. TNF and IL1 circulating levels (mean ± SEM) in 25 normal subjects, 13 first and 24 long-term hemodialysis (HD) patients

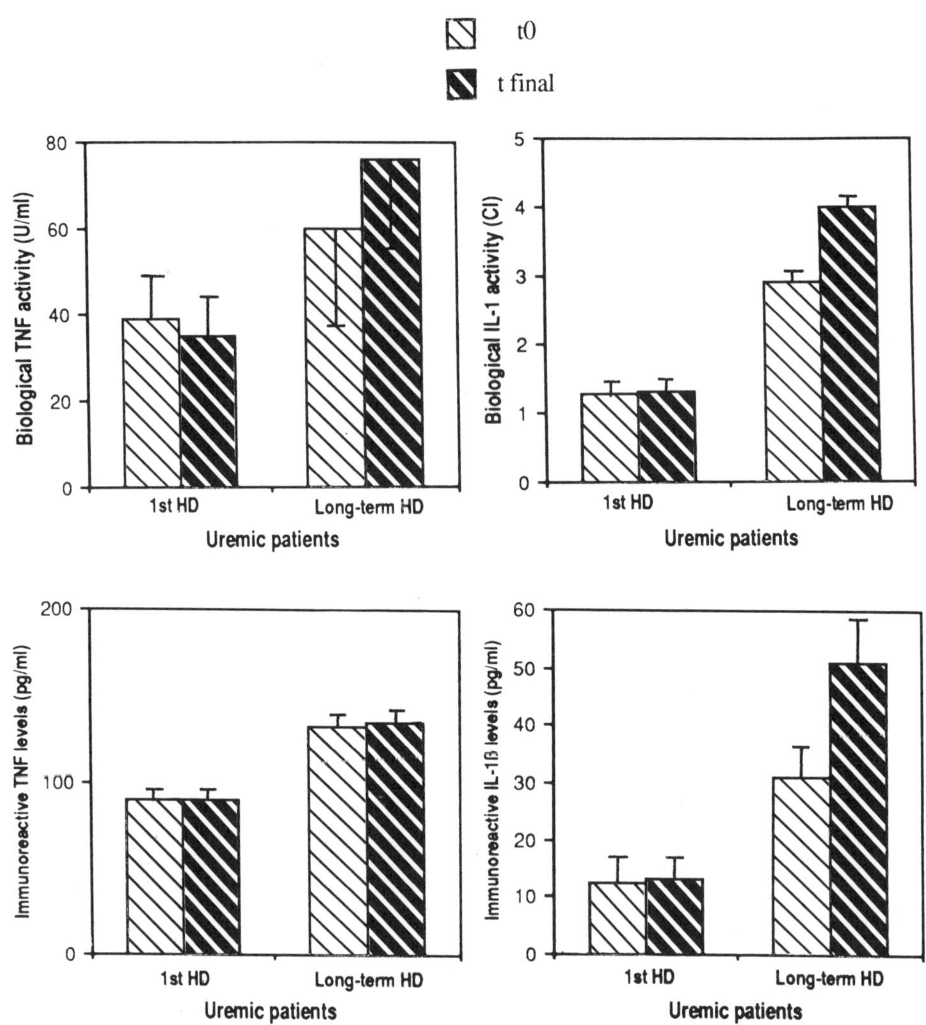

Figure 6. Influence of a single dialysis session on TNF and IL-1 plasma levels determined before t0 and at the end (t final) of the session in 13 first and 22 long term hemodialysis (HD) patients treated with CUP equipped dialysers.

CONCLUSION

Our results support the hypothesis that both dialysis and uremia related factors contribute to modulate ROI and TNF production and that these changes are exacerbated following long-term dialysis treatment. By contrast, IL-1 production is mainly dependent on dialysis and apparently requires recurrence of the dialysis procedure to be significantly increased.

The concomittant presence of increased IL-1 and TNF plasma levels in long-term hemodialysis patients provide further evidence for the chronic occurrence of acute phase changes in these patients. In view of the known synergistic effects of these cytokines it is tempting to propose that their long term production contribute to the genesis of hemodialysis related pathology.

REFERENCES

1. Chatenoud L., Herbelin A., Beaurain G. and Descamps-Latscha B. Adv. Nephrol., 1989, in press
2. Nguyen A.T., Lethias C., Zingraff J., Herbelin A., Naret C. and Descamps-Latscha B. Kidney Int., 1985, *28*, 158
3. Descamps-Latscha B. Contr. Nephrol., 1988, *62*, 132-139
4. Goldfarb B., Nguyen A.T., Landais P., London G., Descamps-Latscha B., Haeffner-Cavaillon N., Jacquot C., Masselot J.P., Kazatchkine M. and Pusineri C. *In* : Abstr. Symp. Bioincompatibilité et Dialyses, Courchevel 1987, p. 42
5. Herbelin A., Nguyen A.T., Zingraff J., Urena P. and Descamps-Latscha B. Kidney Int., 1990, in press
6. Babior B.M. New Engl. J. Med., 1978, *298*, 659-668
7. Nathan C.F. and Cohn Z.A. J. Exp. Med., 1981, *154*, 1539-1553
8. Descamps-Latscha B. *In* : Proc. 4th Int. Symp. Bioluminescence and Chemiluminescence, Freiburg 1986, p. 23 (Wiley, Chichester 1987).
9. Glass G.A., DeLisle D.M., DeTogni P., Gabig T.G., Magee B.H., Markert M. and Babior B.M. J. Biol. Chem. 1986, *261*, 13247-13251
10. Tauber A.I. Blood, 1987, *69*, 711-720
11. Descamps B., Nguyen A.T.and Feuillet-Fieux M.N. Ann. Immunol. (Inst. Pasteur), Paris, 1980, *131D*, 337
12. Descamps-Latscha B., Golub R.M., Nguyen A.T. and Feuillet-Fieux, M.N. J. Immun., 1983, *131*, 2500
13. Goldstein I.M., Roos D., Kaplan H.B. and Weissman G. J. Clin. Invest., 1975, *56*, 1155
14. Allen R., Stjernholm R. and Steele R. Biochem. Biophys. Res. Commun., 1972, *47*, 679-684
15. Allen R.C. Methods Enzymol., 1986, *133*, 449-493
16. Descamps-Latscha B., Nguyen A.T., Golub, R.M. and Feuillet-Fieux M.N. Ann. Immunol. (Inst. Pasteur), 1982, *133C*, 349-364
17. Descamps-Latscha B., Herbelin A., Nguyen A.T., Uzan M. and Zingraff J. Life Support Systems, 1986, *4*, 349
18. Wissow L.S., Greenberg R.S., Burns R.Q., Osofsky S.G., Gutman R.A. and Baker P.J. J. Clin. Immunol., 1981, *1*, 262
19. Cohen M.S., Elliott D.M., Chaplinski T., Pike M.M. and Niedel J.E. Blood, 1982, *60*, 1283
20. Ritchey E.E., Wallin J.D. and Shah S.V. Kidney Int., 1981, *19*, 349-358
21. Tuma S.N., Martin R.R., Mallette L.E. and Eknoyan G. Kidney Int., 1979, *16*, 955
22. Markert M., Waridel P.A., Heierli C. and Wauters J.P. Contr. Nephrol., 1988, *62*, 99-108
23. Beaurain G., Naret C., Marcon L., Grateau G., Drueke T., Urena P., Nelson D.L., Bach J.F. and Chatenoud L. Kidney Int., 1989, *36*, 636-644

24. Petrone W.F., English D.K., Wong K. and Mc Cord J.M. Proc. Natl. Ac ad . Sci. USA, 1979, *77*, 1159
25. Carp H. and Janoff A. J. Clin. Invest., 1979, *63*, 793
26. Horl W.H., Steinhauer H.B. and Schollmeyer P. Kidney Int., 1985, *28*, 791
27. Weiss S.J., Peppin G., Ortiz X., Ragsdale C. and Test S.T. Science, 1985, *227*, 747
28. Del Maestro R.F. Acta Physiol. Scand., 1980 (suppl.) , *492*, 153-168
29. Weber M. and Schmutz J.L. Contr. Nephrol., 1988, *62*, 75-85
30. Memmos D.E., Eastwood J.E., Harris E. and de Wardener H.E. Nephron, 1982, *30*, 106-109
31. Henderson L.W., Koch K.M., Dinarello C.A. and Shaldon S. Blood Purif., 1983, *1*, 3-8
32. Dinarello C.A., Koch K.M. and Shaldon S. Kidney Int., 1988, *33, suppl. 24*, S21-S26
33. Haeffner-Cavaillon N., Fischer E., Bacle F., Carreno M.P., Maillet F., Cavaillon J.M. and Kazatchkine M.D. Contrib. Nephrol., 1988, *62*, 86-98
34. Bingel M., Lonnemann G., Koch K.M., Dinarello C.A. and Shaldon S. Lancet, 1987, *1*, 14
35. Luger Z., Kovarik J., Stummvoll H.K., Urbanska A. and Luger T.A. Kidney Int., 1988, *32*, 84-88
36. Haeffner-Cavaillon N., Cavaillon J.M., Ciancioni C., Bacle F., Delons S. and Kazatchkine M.D. Kidney Int., 1989, *35*, 1212-1218
37. Dinarello C.A. Immunol. Lett., 1987, *16*, 227-232
38. Le J. and Vilcek J. Lab. Invest., 1987, *56*, 234-248

PHAGOCYTE FUNCTION IN UREMIA

R. Vanholder, R. Dell'Aquila and S. Ringoir

Nephrology Department
University Hospital
Ghent, Belgium

Several studies in uremic patients have demonstrated an increased
susceptibility to infection (1-3), and an enhanced risk for cancer
(4-5). Antibody formation in response to vaccination (e.g. hepatitis B
or influenza) is suppressed (6-7). Skin anergy develops and immune
diseases burn out during the progress of renal failure. We may assume an
acquired immune deficiency in these patients. In this suppression of the
immune system, a central role may be played by phagocyte cells. One of
their main functions is the ingestion and killing of bacteria, but these
cells are also secretors of humoral factors, with regulatory and stimulatory effects on lymphocytes and ohter cells of the immune system.
Several of these humoral factors (e.g. cytokines, leukotrienes) have
only recently been detected and may be of major importance in the fine-
tuning of the reaction of the organism towards foreign material (8). A
remarkable aspect is the limited degree of specialisation of the phagocytic reaction : once stimulated, a virtually identical chain of events
takes place, whatever the stimulus. It consists of the uptake of the
challenging microparticle, if any, the activation of a series of metabolic reactions defined as the respiratory burst and the production of
oxygen free radicals that are lethal for bacteria (9-10). Once the phagocyte is activated, these free radicals are released as an endpoint, by
the way not only killing bacteria, but also being a potential cause of
damage to the normal tissues and/or of organ dysfunction. There are
several mechanisms that may be held responsible for a suppression of
phagocytic function in renal failure. The uremic syndrome results in an
accumulation of solutes with potential toxic effects on phagocytic function (table I). A negative effect may also result from the uremic accumulation of free radical scavangers, the latter compounds being
responsible for the neutralization of free radicals as end products of
the phagocytic process. Uremia is also characterized by deficiencies,
e.g. of zinc or fibronectin, that may result in inadequate phagocytic
function. Metabolic dysfunctions at least partially related to uremic
anemia, but also to other metabolic disturbances such as glucose metabolism, may play a further role. Finally, different forms of renal replacement therapy may interfere with phagocytic function : certain types of
hemodialysis may cause bioincompatibility related inflammation, e.g. as
a consequence of complement or leukocyte activation, or as a result of
endotoxin transfer from dialysate to the blood compartment : it is
known from other conditions that such a chronic inflammation may cause
phagocytic dysfunction; rejection of renal transplants is prevented by

TABLE I . potential causes of phagocyte dysfunction in uremia

- Toxin accumulation
    Polyamines, spermine
    Endorphins
    Guanidino compounds

- Accumulation of free radical scavangers
    Uric acid
    Indoles
    Phenols

- Deficiencies
    Zinc
    Fibronectin

- Metabolic dysfunction
    Glucose metabolism
    Protein synthesis
    Oxygen transport

- Bioincompatibility of renal replacement therapy
    Hemodialysis : complement and leukocyte activa-
    tion (cuprophan)
    CAPD : effect of dialysate on peritoneal
    macrophages
    Transplantation : immune suppression

immune suppressive agents, that in turn suppress phagocytic reactivity; the intra-peritoneal administration of dialysate in CAPD may deteriorate the usual peritoneal defense mechanisms. In spite of these attractive theoretical hypotheses suggesting several forms of immune suppression in uremia, the available scientific data are less self-evident; although phagocytic function in uremia has been the subject of repeated studies, the available literature does not allow any consistent conclusion, different authors pointing to suppression, unaltered function or even stimulation of the phagocyte system depending on the study conditions. In the present paper, the current literature data concerning phagocyte function in uremia will be reviewed, underlining the frequent contradictions in the results obtained, and attempting to discern some of the reasons for these discrepancies. In tables II and III currently cited papers on phagocyte function in uremia are reviewed. They are subdivided in studies on uremia as such (mostly on samples of non-dialysed uremic patients and pre-dialysis samples of hemodialysed patients) (table II), and studies specifically focussing on intra-dialytic evolution (table III). The results are obviously divergent. Some of these discrepancies are due to methodological and/or technical factors; we thought that it would be interesting to try to understand some of the factors that might interfere here.

PATIENT POPULATIONS

One potential bias may be related to the composition of the patient populations under study. From table II, it appears that of 19 studies considering end-stage renal failure patients or pre-dialysis samples of hemodialysed patients, 12 are limited to one specific

TABLE II . Characteristics of studies considering phagocyte function in uremia (pre-terminal renal failure or pre-dialysis samples)

Author (Ref)	Patient population	n	Artificial medium	Ficoll Hypaque	Staph. aureus E.Coli/Candida	CL	Effect
Hosking et al,1976 (11)	HD	22	+	+	+	−	=
Abrutyn et al,1977 (12)	ESRF	11	+	−	+	−	=
Björksten et al,1978 (13)	HD	34	+	−	−	−	−
Hallgren et al, 1979 (14)	HD	37	+	?	−	−	−
Lespier-Dexter et al,1979 (15)	ESRF/HD	8/15	−	−	−	−	−/=
Sutowicz et al,1979 (16)	ESRF/HD	10/14	−	−	−	+	=
Wardle et al, 1980 (17)	ESRF	8	+	−	−	+	+
Ritchey et al,1981 (18)	HD	13	+	−	−−	+	−
Tuma et al,1981 (19)	ESRF+HD	7/5	+	+	−	+	=/+
Nelson et al,1983 (20)	Animals	<20	+	−	+	−	=
Wierusz-Wysocka et al,1983 (21)	HD	12	+	?	−	−	−
Glazer et al,1984 (22)	HD	15	+	+	−	+	−
Nguyen et al,1985 (23)	HD	35	−	−	−	+	−/=
Flament et al,1986 (24)	HD	21/19	−	−−	−	+	−
Rhee et al,1986 (25)	ESRF/CAPD/HD	44/17/18	−	−	+	+	+
Cantinieaux et al,1988 (26)	HD	9	+	+	+	−	−
Lewis et al,1988 (27)	CAPD/HD	16/15	+	+	−	+	−/=
Hirabayashi et al,1988 (28)	ESRF/CAPD/HD	8/5/33	+	+	−	+	−/=
Lucchi et al,1989 (29)	ESRF/HD	27/33	+	+	−	+	−

HD : hemodialysis; CAPD : chronic ambulatory peritoneal dialysis, ESRF : end stage renal failure (non-dialysed)
CL : Chemiluminescence method or other method estimating free radical production
n : number of patients under consideration.
Effect : phagocyte function enhanced (+), unaltered (=) or suppressed (−) vs normal

TABLE III . Characteristics of studies considering intradialytic phagocytic function

Author (Ref)	Dialyzer membrane	n	Artificial medium	Ficoll Hypaque	Staph.areus/ E.Coli/Candida	CL	Effect
Henderson et al,1975* (30)	C/NC	3	+	−	+	−	−/=
Mac Gregor et al,1977 (31)	C	−	−	−	−	−	+
Lespier-Dexter et al,1979 (15)	C	3	−	−	−	−	+
Klempner et al,1980 (32)	C	3	+	+	−	−	−
Wissow et al,1981 (33)	C	7	+	−	+	+	−
Cohen et al,1982 (34)	C	16	+	+	−	+	−
Spagnuolo et al,1982 (35)	C	12	+	+	−	−	=
Bauer et al,1983 (36)	C/NC	5	−	−	+	+	−/=
Wierusz-Wysocka et al,1983 (21)	C	12	+	?	−	−	−
Nguyen et al,1985 (23)	C/NC	13/22	−	−	−	+	−/=
Markert et al,1988 (37)	C/NC	12	+	−	−	+	−/=
Rocattello et al,1989* (38)	C/NC	4−6	+	−	−	+	−

* : In vitro, with blood cells of healthy subjects
C : cellulosic, NC : synthetic non-cellulosic
Effect : phagocytic function enhanced (+), unaltered (=) or suppressed (−) vs normal.

patient group (e.g. only non-dialysed patients). Even when considering
several patient groups together, longitudinal evolution has rarely
been evaluated, i.e. pre-dialysis end stage renal failure or hemodialysis treated patients were considered all together as a general
group, not evaluating the longitudinal evolution per patient and the
different stages of the disease, nor the degree of renal failure or
the time spent on dialysis. Table III summarizes the data, obtained
from the evaluation of intradialytic samples. In several studies
(7/12) only dialyzers containing cellulosic cuprophan membranes are
evaluated, and not synthetic non-cellulosic dialyzer membranes. The
first are known to cause complement and leukocyte activation, so that
phagocytic response during treatment with these dialyzers might be
different from the one obtained during treatment with synthetic dialyzers, that provoke no such inflammatory response. Two of the studies
(30, 38) are performed in an entirely artificial medium, submitting
separated white blood cells of healthy subjects to in vitro treatment.

PATIENT NUMBERS

The mean number of patients evaluated per study is 18 for non dialy-
tic samples and 9 for intra-dialytic studies. Of the 19 studies considering the uremic status (end stage renal failure before the start of
dialysis and dialysed patients on pre-dialysis samples), 10 are based on
15 samples or less. Of the 12 studies evaluating intra-dialytic evolution only 2 are based on 15 samples or more. This may be a major draw-
back, as data on phagocytic function are often subjected to an important
scatter, so that the absence of significant differences has no objective
meaning, in studies on groups of such a limited extent.   Also, evaluating small patient groups, may result in selection of specific
patient subgroups, again giving rise to unreliable conclusions.

LEUKOCYTE ISOLATION

The question whether leukocytes should be isolated or not before the
evaluation of their reactivity in this type of study, remains a matter
of debate. A practical consideration is that evaluations on isolated
leukocytes are more complex, labor-intensive and timeconsuming, and that
larger quantities of blood are needed in patients that are often anemic.
All these data may discourage the evaluation of large patient groups.
The question arises to what extent the leukocyte population obtained
after several steps of isolation and manipulation is representative for
the leukocytes that were present at the start of the rocedure, before
any separation. Especially, the use of Ficoll-Hypaque gradients has been
claimed to cause a selection of less virulent phagocytes (38-40). Other
isolation methods have recently also been demonstrated to cause alterations in phagocyte function, mainly activation of their metabolism (41).
Considering the data of table II and III, 14/9 and 8/12 studies used
isolated leukocytes that were maintained in an artificial medium.
Amongst them, 7/14 and 3/8 were based on a Ficoll-Hypaque gradient.
Although we are convinced that the study of isolated leukocytes results
in useful basic information, the potential bias induced by this approach
should thus be taken into account. It might therefore be useful to consider techniques that allow an immediate evaluation of blood samples,
with leukocytes in their natural environment, at or immediately after
the moment of their collection, rather than going through a series of
complex separation steps.

CHALLENGING MICROPARTICLES

For the study of phagocytic function in uremia, artificial stimuli,
e.g. latex or zymosan, have mostly been used. Else, soluble non-particle
stimuli, like phorbol esters or fMLP are used. These challenges might
not entirely be representative for what happens when phagocytes are
confronted with a real bacterial challenge during infection. Latex e.g.
can be considered as a microparticle as such. Zymosan causes a more
complex set of events, associating particle ingestion to receptor
mediated activation (e.g. mannose receptors), and complement activation.
Phorbol esters, like phorbol myristate acetate (PMA) induce a direct
activation of protein kinase C without intermediate steps. Finally,
fMLP, as another soluble compound and a chemotactic peptide, induces
receptor mediated stimulation, by a pathway involving intracellular
calcium shifts (8). According to table II and III, studies using bacteria or other microbes, which could be considered as a more natural
type of stimulus, are extremely scant (5/19 and 3/12 respectively).

EVALUATION OF PHAGOCYTIC ACTIVATION

Most of the studies considered, evaluate only one single aspect of
phagocytic function. From table II and III, it is clear that 11/19 studies not considering intra-dialytic data, and 6/12 studies considering
intra-dialytic data, are based on the estimation of free-radical production, either by determination of chemiluminescence production or by
other related tests (e.g. reduction of nitrobluetetrazolium). The
correct interpretation of free radical production in uremia may be rendered inaccurate by the accumulation of free radical scavangers in serum
and blood cells of renal failure patients. Current uremic toxins, such
as uric acid, phenols and indoles, are strong scavangers (17, 42-43).
Although the exact influence of their retention on the final outcome of
tests of phagocyte function is uncertain, it is conceivable that their
presence in cellular cytoplasm or in the serum surrounding cells will
influence the final interpretation of phagocytic function when based on
estimates of free radical production, even when the basic metabolic
activity of phagocytes remains unaltered. This may bias the eventual
conclusions, especially since the concentration of these scavangers in
the serum is influenced by the degree of renal failure and the elimination pattern of different dialysis strategies used.

In whole blood samples, the scavanging capacity of red blood cells
may cause an additional bias; this may be an important pitfall, especially in view of the marked differences in erythrocyte counts that may
occur amongst patients, related to their degree of renal failure and
their basic renal disorder. The use of estimation of chemiluminescence
production may further be hampered by the necessity to dilute the
samples ten- to one hundredfold, hereby not maintmaining the phagocytes
in their original milieu. Therefore, we think that methods other than
those based on estimations of free radical production, may offer supple-
mentary information for the study of phagocytic function in uremia,
although these methods have rarely been used.

GENERAL CONCLUSIONS ON THE AVAILABLE LITERATURE

Characteristics that may have a negative influence on the final
interpretation of studies on phagocytic function in uremia are :
1) the composition of the patient groups;
2) the lack of longitudinal data on the evolution during the
progress of the syndrome;

3) the absence of a sufficient number of patients under study;
4) the potential changes in phagocyte function due to the use of
separation techniques resulting in phagocyte populations that are no
more present in their natural environment;
5) the additional selection resulting from Ficoll-Hypaque separation
methods;
6) the lack of challenging studies with natural stimuli, such as
living bacteria;
7) the use of studies based on the estimation of superoxide generation in the presence of free oxygen radicals that are currently
retained in uremia;
8) the lack of evaluations with different dialyzer membrane types,
during intra-dialytic studies;
9) the performance of studies in an in vivo setting, using separated
white blood cells of healthy donors that may be irrepresentative for
the uremic state;
10) the lack of studies using more than one evaluation method.

Considering all studies reported in table II and III, most studies
have one or more of the above mentioned characteristics (table IV).

RESULTS OBTAINED BY OUR OWN GROUP

Most studies in our unit estimating phagocyte function, were performed by an estimation of the quantity of radioactive CO_2 produced from
labelled glucose-1-C during the r.spiratory burst by the hexose-
monophosphate shunt (fig. 1) (44-47). This test method can be used on
whole blood samples, allows the processing of a large number of samples,
and is not directly estimating superoxide production. Different
challenging particles or humoral substances can be used, for our studies
most currently latex and zymosan, but also staphylococcus aureus, phorbol myristate acetate and fMLP. The quantity of CO_2-production is
related to the degree of activation of the hexose monophosphate shunt.
The energy produced during this process is used for the activation of
an enzyme system called NAD(P)H-oxidase, resulting in the production
of superoxide, that is transformed in even more toxic free radicals,
such as hydrogen peroxide, hypochlorous acid, singlet oxygen and
hydroxyl radical (fig. 1). The latter compounds are used in the
killing process of bacteria. This test

TABLE IV . Phagocyte function in uremia : potential biasses

- Small patient groups	16/28[*]	(57 %)
- Ficoll Hypaque	10/26	(39 %)
- Leukocytes in heteroligous environment	22/28	(79 %)
- No bacterial stimuli	21/28	(75 %)
- Measurement free radical production	16/28	(57 %)

Number of biasses per study :		
1	2	studies
2	10	studies
3	4	studies
4	11	studies
5	1	study

Summary of tables II and III - [*] Illustrated is the relative
number of studies biassed for the given factor

method gives a direct estimation of metabolic activity of phagocyte
cells, but was also chosen because of its suppression in chronic granulomatous disease, a pediatric disorder, that like uremia, is mainly
characterized by a susceptibility for gram-negative and staphylococcal
infections (48). Earlier studies of our group, using this test method,
had revealed that :
1) phagocyte function in uremia was suppressed (44);
2) a factor in urine and in ultrafiltrate of uremic serum suppressed
function of healthy phagocytes (45-46);
3) uremic phagocyte suppression could at least in part be corrected
by the administration of the new antibiotic cefodizime (44);
4) an activation of basic phagocyte activity during cuprophan dialysis is absent during treatment with other, synthetic membranes (47).

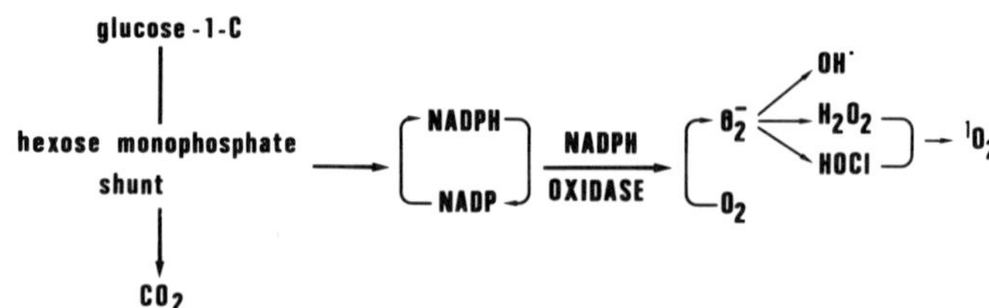

Fig. 1. Flow chart of the biochemical reactions taking place
during the respiratory burst of phagocytes, resulting in
free radical production.

Preliminary results of more elaborate studies demonstrate a normal
phagocyte metabolic function up to a creatininemia of 6 mg/100 ml (Ccrea
= 15 ml/min), followed by a period of gradual suppression when renal
function further deteriorates ; the start of hemodialysis causes a
further fall in phagocyte function, to reach a nadir 3 to 4 weeks after
the start of dialysis. This nadir is more pronounced with cuprophan
dialyzers (49). Phagocyte metabolic function gradually improves when
dialysis is prolonged, but a normalization is only observed after $\pm$ 10
years of treatment. The use of erythropoietin corrects phagocyte dysfunction. Washing of healthy blood cells with uremic serum causes a
suppression of reactivity (fig. 2), at least when cells from healthy
persons are used with a sufficient basic reactivity. These data were
mainly obtained with latex and zymosan as challenging microparticles,
but were confirmed by use of staphylococcus aureus and phorbol
myristate. They were also confirmed by parallel studies measuring
intracellular free radical production by flow cytometric analysis (49).
Separate intradialytic studies, indicate that there is an intra-dialytic
lack of phagocyte reactivity versus stimuli, that is absent during
treatment with non-cellulosic membranes, but also with hemophan and
reused cuprophan. Our data indicate that phagocyte metabolic dysfunction
is present during uremia, due to a metabolic defect, that involves glucose metabolism and as suggested by the PMA-data, protein kinease C. It
improves with a correction of uremic anemia. Disturbances are accen-
tuated by the bioincompatibility versus leukocytes of dialysis membranes, not only during dialysis, but also in pre-dialysis samples,

especially during the first weeks after the start of dialysis. The
progress of dysfunction in pre end-stage renal failure, suggests also
a toxic involvement, but it remains difficult to demonstrate
suppression, when washing healthy cells with uremic serum.

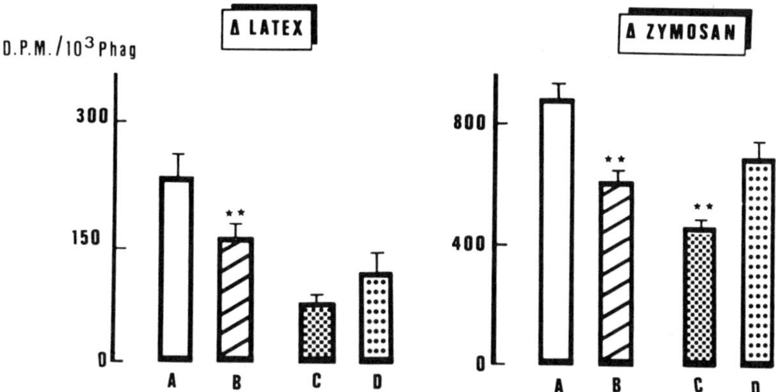

Fig.2   Effect of washing of healthy blood cells with uremic
        serum (A vs B) and uremic blood cells with healthy
        serum (C vs D) on $^{14}CO_2$-production during respiratory
        burst activation. Washing of healthy blood cells with
        uremic serum causes a suppression of metabolic activity,
        whereas washing of uremic blood cells with normal serum
        causes a partial correction of metabolic dysfunction.

CONCLUSIONS

    Literature data on the evolution of phagocytic function are incon-
sistent, which can mainly be attributed to differences in methodological
approach. It is conceivable, however, that phagocyte function, and
especially their ability to procduce energy for the transformation of
oxygen in free radicals, is hampered at least at some stages of the
disease. According to our own data, this is especially the case during
the months preceding the start of dialysis, and in the period immediately following this start of dialysis. Changes are related to the
bioincompatibility of renal replacement therapy, toxin retention and
metabolic dysfunction. The exact knowledge on the factors influencing
uremic phagocyte function can be of major help in the prevention of
infections and other complications in renal patients, either or not
treated by renal replacement therapy.

REFERENCES

1. D. Kleinknecht, P. Jungers, J. Chanard, C. Barbanel, and D.
    Ganeval, Uremic and non-uremic complications in acute renal
    failure : Evaluation of early and frequent dialysis on progno-
    sis, Kidney Int. 1:190 (1972).

2. JZ. Montgomerie, G.M. Kalmanson, and L.B. Guze, Renal failure,
    and infection, Medicine. 47:1 (1968).

3. M.C. Belcon, E.K.M. Smith, L.M. Kahana, and A.F. Shimuzu,
    Tuberculosis in dialysis patients, Clin. Nephrol. 17:14(1982).

4. A. Lindner, V.T. Farewell, and D.J. Sherrard, High incidence of
    neoplasia in uremic patients receiving long-term dialysis,
    Nephron. 27:292(1981).

5. F.K. Port, N.E. Ragheb, A.G. Schwartz, and V.M. Hawthorne,
    Neoplasms in dialysis patients : A population-based study, Am.
    J. Kidney Dis. 14:119 (1989).

6. P. Rautenberg, I. Teifke, T. Schlegelberger, and U. Ullmann,
    Influenza subtype-specific IgA and IgG responses in patients on
    hemodialysis after influenza vaccination, Infection.
    16:323 (1988).

7. R.W. Steketee, M.E. Ziarnik, and J.P. Davis, Seroresponse to
    hepatitis B vaccine in patients and staff of renal dialysis
    centers, Wisconsin, Am. J. Epidemiol., 127:772 (1988).

8. J.D. Lambeth, Activation of the respiratory burst oxidase in
    neutrophils : on the role of membrane-derived second
    messengers, Ca^{++}, and protein kinase C, J. Bioenerg.
    Biomembranes, 20:709 (1988).

9. J.A. Badwey, and M.L. Karnovsky, Active oxygen species and the
    functions of phagocytic leukocytes, Am. Rev. Biochem.
    49:695 (1980).

10. M.A. Foy, and L. Simchowitz, Recent developments in leukocyte
    research. In: The Year in Immunology. Cellular, Molecular and
    Clinical Aspects. J.M. Cruse, R.E. Lewis, eds., Karger, Basel,
    208 (1989).

11. C.S. Hosking, R.C. Atkins, D.F. Scott, S.R. Holdsworth,
    M.G. Fitzgerald, and M.J. Shelton, Immune and phagocytic func-
    tion in patients on maintenance dialysis and post-
    transplantation, Clin. Nephrol. 6:501 (1976).

12. E. Abrutyn, N.W. Solomons, L. St. Clair, R.R. MacGregor, and
    R.K. Root, Granulocyte function in patients with chronic renal
    failure: surface adherence, phagocytosis, and bacterial acti-
    vity in vitro, J. Infect. Dis. 135:1 (1977).

13. B. Björksten, S.M. Mauer, E.L. Mills, and P.G. Quie, The effect
    of hemodialysis on neutrophil chemotactic responsiveness, Acta
    Med. Scand. 203;67 (1978).

14. R. Hällgren, K.E. Fjellström, L. Hakansson, and P. Venge, Kinetic
    studies of phagocytosis. II. The serum-independent uptake of
    IgG-coated particles by polymorphonuclear leukocytes from ure-

mic patients on regular dialysis treatment, J. Lab. Clin. Med. 94: 277 (1979).

15. L.E. Lespier-Dexter, C. Guerra, W. Ojeda, and M. Martinez-Maldonado, Granulocyte adherence in uremia and hemo-dialysis, Nephron. 24:64 (1979).

16. W. Sutowicz, Z. Komorowska, Z. Hanicki, T. Cichocki, and O. Smolenski, The enzymatic activities and NBT reduction test of granulocytes in untreated and dialysed uraemic patients, Acta Haemat. 61:144 (1979).

17. E.N. Wardle, and R. Williams, Polymorph leucocyte function uraemia and jaundice, Acta Haemat. 64:157 (1980).

18. E.E. Ritchey, J.D. Wallin, and S.V. Shah, Chemiluminescence and superoxide anion production by leukocytes from chronic hemo-dialysis patients, Kidney Int. 19:349 (1981).

19. S.N Tuma, R.R. Martin, L.E. Mallette, and G. Eknoyan, Augmented polymorphonuclear chemiluminescence in patients with secondary hyperparathyroidism, J. Lab. Clin. Med. 97:291 (1981).

20. J. Nelson, D.J. Ormrod, and T.E. Miller, Host immune status in uremia. IV. Phagocytosis and inflammatory response in vivo, Kidney Int. 23:312 (1983).

21. B. Wierusz-Wysocka, H. Wysocki, R. Czarnecki, H. Siekierka, K. Baczyk, and K. Wysocki, Influence of hemodialysis on plasma che-motactic activity and the chemotactic responsiveness of poly-morphonuclear neutrophils, Artif. Organs. 7:159 (1983).

22. T. Glazer, P. Fishman, B. Klein, J. Levi, and M. Djaldetti, Generation of superoxide anions during phagocytosis by monocy-tes of uremic patients, Nephron. 38:40 (1984).

23. A.T. Nguyen, C. Lethias, J. Zingraff, A. Herbelin, C. Naret, and B. Descamps-Latscha, Hemodialysis membrane-induced activation of phagocyte oxidative metabolism detected in vivo and in vitro within microamounts of whole blood, Kidney Int. 28:158 (1985).

24. J. Flament, M. Goldman, Y. Waterlot, E. Dupont, J. Wybran, and J.L. Vanherweghem, Impairment of phagocyte oxidative metabolism in hemodialyzed patients with iron overload, Clin. Nephrol. 25:227 (1986).

25. M.S. Rhee, M.D. Mc.Goldrick, and H.J. Meuwissen, Serum factors from patients with chronic renal failure enhances polymorpho-nuclear leukocyte oxidative metabolism, Nephron. 42:6 (1986).

26. B. Cantinieaux, J. Boelaert, C. Hariga, and P. Fondu, Impaired neutrophil defense against Yersinia enterocolitica in patients with iron overload who are undergoing dialysis, J. Lab. Clin. Med. 111:524 (1988).

27. S.L. Lewis, D.E. Van Epps, and D.E. Chenoweth, Alterations in chemotactic factor-induced responses of neutrophils and monocy-tes from chronic dialysis patients, Clin. Nephrol. 30:63 (1988).

28. Y. Hirabayashi, T. Kobayashi, A. Nishikawa, H. Okazaki, T. Aoki, J. Takaya, and Y. Kobayashi, Oxidative metabolism and phagocytosis of polymorphonuclear leukocytes in patients with chronic renal failure, Nephron. 49:305 (1988).

29. L. Lucchi, G. Cappelli, M.A. Acerbi, A. Spattini, and E. Lusvarghi, Oxidative metabolism of polymorphonuclear leukocytes and serum opsonic activity in chronic renal failure, Nephron. 51:44 (1989).

30. L.W. Henderson, M.E Miller, R. W. Hamilton, and M.E. Norman, Hemodialysis leukopenia and polymorph random mobility - a possible correlation, J. Lab. Clin. Med. 85:191 (1975).

31. R.R. MacGregor, Granulocyte adherence changes induced by hemodialysis, endotoxin, epinephrine and glucocorticoids, Ann. Intern. Med. 86:35 (1977).

32. M.S. Klempner, J.I. Gallin, J.E. Balow, and D.P. Van Kammen, The effect of hemodialysis and C5ades arg on neutrophil subpopulations, Blood 55:777 (1980).

33. L.S. Wissow, R.S. Greenberg, R. O. Burns, S.G. Osofsky, R.A. Gutman, and P.J. Baker, Altered leukocyte chemiluminescence during hemodialysis, J. Clin. Immunol. 1:262 (1981).

34. M.S. Cohen, D.M. Elliott, T. Chaplinski, M.M. Pike, and J.E. Niedel, A defect in the oxidative metabolism of human polymorphonuclear leukocytes that remain in circulation early in hemodialysis, Blood 6:1283 (1982).

35. P.J. Spagnuolo, S.H. Bass, M.C. Smith, K. Danviriyasup, and M.J. Dunn, Neutrophil adhesiveness during prostacyclin and heparin hemodialysis, Blood 60:924 (1982).

36. H. Bauer, H. Brunner, H.F. Franz, and B. Bultmann, Leucocyte function tests during hemodialysis with different dialysis membranes, Contr. Nephrol. 36:9 (1983).

37. M. Markert, C. Heierli, T. Kuwahara, J. Frei, and J.P. Wauters, Dialyzed polymorphonuclear neutrophil oxidative metabolism during dialysis : a comparative study with 5 new and reused membranes, Clin. Nephrol. 29:129 (1988).

38. D. Roccatello, G. Mazzucco, R. Coppo, G. Piccoli, C. Rollino, B. Scalzo, M.G. Guerra, G. Cavalli, O. Giachino, A. Amore, F. Malavasi, and L.M. Sena, Functional changes of monocytes due to dialysis membranes, Kidney Int. 35:622 (1989).

39. S.L. Lewis, and D.E. Van Epps, Neutrophil and monocyte alterations in chronic dialysis patients, Am. J. Kidney Dis. 9:381 (1987).

40. S.L. Lewis, D.E. Van Epps, and D.E. Chenoweth D.E., Density changes in leukocytes following hemodialysis exposure to chemotactic factors, Am. J. Nephrol. 6:34 (1986).

41. Smith D.M., Hanes B., J.A. Johnson, and R.A Turner, Functional and metabolic heterogeneity among normal neutrophil subpopulations, J. Clin. Lab. Anal. 3:174 (1989).

42. E.N. Wardle, Chemiluminescence and superoxide anions generated
     by phagocytes in uraemia, Nephron 40:379 (1985) (letter).

43. K.J. Maples, and R.P. Mason, Free radical metabolite of uric
     acid, J. Biol. Chem. 263:1709 (1988).

44. R. Vanholder, N. Van Landschoot, E. Dagrosa, and S. Ringoir,
     Cefodizime : a new cephalosporin with apparent immune-
     stimulating properties in chronic renal failure, Nephrol.
     Dial. Transplantation. 2:221 (1988).

45. S. Ringoir, N. Van Landschoot, and R. De Smet, Influence of
     human urine on phagocytosis. U.C. Dubach and U. Schmidt, eds.,
     Hans Huber Verlag, Bern, 358 (1979).

46. S.M.G. Ringoir, N. Van Landschoot, and R. De Smet, Inhibition
     of phagocytosis by a middle molecular fraction from ultra-
     filtrate, Clin. Nephrol. 13:109 (1980).

47. R.C. Vanholder, A. Dhondt, and S.M.G. Ringoir, Challenge of
     phagocyte metabolism by extracorporeal circulation and membrane
     contact, Trans. Am. Soc. Artif. Intern. Organs. 34:543 (1988).

48. J.I. Gallin, E.S. Buescher, B.E. Seligmann, N. Nath, T.
     Gaither, and P. Katz, Recent advances in chronic granulomatous
     disease, Ann. Intern. Med. 99:657 (1983).

49. R. Vanholder, S. Ringoir, and R. Hakim, Phagocytic function in
     uremic and hemodialysed patients. To be published.

NEW ASPECTS ON OXIDATIVE METABOLISM OF NEUTROPHILS DURING

HEMODIALYSIS ON DIFFERENT DIALYZER MEMBRANES

M. Markert[1] and J.P. Wauters[2]

Laboratory of Clinical Chemistry[1]
and Division of Nephrology[2],
University Hospital, CH-1011 Lausanne
Switzerland

INTRODUCTION

The marked neutropenia observed during the initial phase
of haemodialysis (HD) with cellulosic membranes results from
pulmonary sequestration of neutrophils[1]. The mediators responsible for this phenomenom have not been completely established, but complement activation leading to the formation
of the cleavage products C_{3a} and C_{5a} during the first 15 min
of HD has been devoted the most important or exclusive part
(1,2). However other mechanisms, such as direct cell-membrane
interaction, cellular responses or protein adsorbtion on membranes could also be involved. Indeed, the contact of the
blood with various types of dialyzer membrane leads to a degree of leukopenia not always linked to the degree of complement activation[3] nor to the transient alterations of oxidative metabolism[4] and other neutrophil functions[5] observed during haemodialysis.

Our study was further devoted to a better understanding
of this HD-induced neutropenia in relation to the oxidative
metabolism of neutrophils in vivo and in vitro.

METHODS

Patients and dialyzers

Twelve-sixteen patients entered these studies utilyzing
cuprophan (CU), cellulose acetate (CA), polycarbonate (PC),
polysulfone (PS) and polyacrilonitrile (PAN) dialyzer membranes as already described[3,6].

Neutrophil purification

Purified preparation of polymorphonuclear leukocytes
were obtained in a single centrifugation step on a discontinuous Percoll gradient (76%/60%)[3].

Oxygen radical determination

Luminol and lucigenin-enhanced chemiluminescence (CL) were measured according to a published procedure[4]. Hydrogen peroxide was measured as described with slight modifications[7].

Membrane fragments, namely cuprophan (CU), cellulose acetate (CA), polycarbonate (PC), polysulfone (PS), polymethylmethacrylate (PMMA) and polyacrilonitrile (PAN) were washed 3 times with saline prior to use. The saline washed membranes were also further treated with fresh or heat-unactivated plasma for 10 min at 37 °C, then washed 3-4 times with saline.

Tubes containing HBSS buffer, 20 mg of the dialyzer membrane fragments, 2 mmol.l^{-1} NaN$_3$, 2 x 10^6 PMN, in a final volume of 1 ml were incubated for 20 min at 37 °C in a shaking water bath. The reaction was stopped by centrifuging at 1200 g for 5 min at 4 °C. 500 μl of supernatant were transferred into fresh tubes containing phenol red (0,01 g.l^{-1}), horse-radish peroxidase (HRP) (Boehringer, Mannheim, West-Germany) (50 ng.ml^{-1}) and HBSS to a final volume of 1 ml. The reaction mixture was incubated for 5 min at room temperature, after which 10 μl of 1 mol.l^{-1} NaOH were added. The absorbance was measured spectrophotometrically at 610 nm. A standard curve with known amounts of H$_2$O$_2$ and containing the same amount of NaN$_3$ as the sample was run simultaneously.

Complement measurement

Concentrations of C$_{3a\ des\ arg}$ were measured with a radioimmunoassay (Hans Rhan and Co., Amersham International, Zürich).

Protein adsorption

A capillary cuprophan dialyzer was rinsed with saline, primed at room temperature for 10 min with human albumin (HA), immunoglobulins (Ig) or total plasma proteins (PP) and rinsed before starting a clinical dialysis[8].

RESULTS AND DISCUSSION

Oxidative metabolism of neutrophils during HD

The kinetic of oxygen radical production of cells during HD on cuprophan (CU) as measured by luminol-amplified CL is shown in fig.1. Cells purified from a control, cells taken before dialysis and cells taken at 5 and 60 min of HD and stimulated with opsonized zymosan showed the same pattern of radical production, whereas cells taken at 15 min of HD displayed a significant lower response. This defective functional capacity was found to be membrane dependent[4] and was attributed either to systemic down-regulation of the cells by activated complement, or to the removal of most active cells leading to a globally defective population of cells in circulation at maximum neutropenia. When the responses of time 0 dialyzed cells were compared to control cells, no difference

was found indicating no cellular defect of cells before HD.
Yet the opsonizing capacity of time 0 dialyzed plasma was
found to be decreased before and further during HD[4] which
could at least in part explain the increased susceptibility
to infections of dialyzed patients.

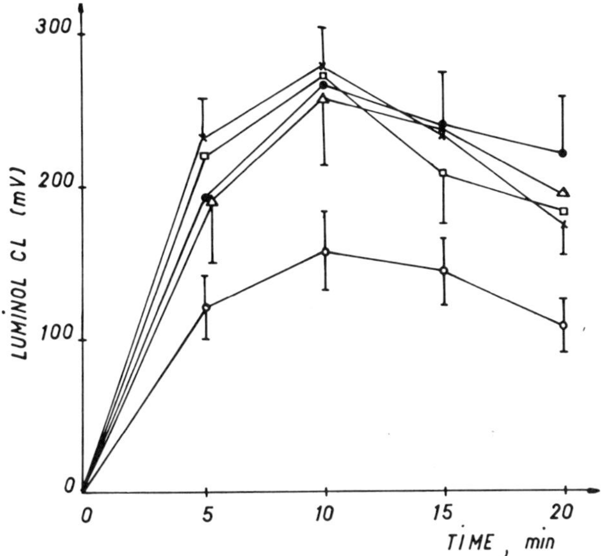

Fig.1.   Kinetic of oxygen radical production by cells
           collected during hemodialysis on CU at time 0
           ———●———, time 5 min ——x——, time 15 min ——o——,
           time 60 in ——□—— and control cells —— △ ——.
           Values are the mean ± SEM of 12 experiments.

In vitro effect of membrane contact

H_2O_2 production by neutrophils in contact with various
dialyzer membranes fragments is depicted in fig.2. All the
membranes were able to stimulate H_2O_2 production above control level. It is worth noting that the highest response was
observed with PAN, a membrane known for its good biocompatibility properties. Upon coating the membranes with plasma, a
dual action on membrane-induced oxygen radical production
could be observed, that is highly stimulatory for CU, and
inhibitory for PAN and PMMA. After treatment with heat-
inactivated plasma only CU and PS were less able to activate
cells.

These results suggest that neutrophil activation is modulated by adsorbed plasma factors related to complement or not as also shown when measuring other oxygen radicals[9].

To further analyze which plasma component could be involved in modulating these effects, albumin-coated membranes were tested (table 1).

Compared to naked membranes or to plasma-treated membranes, coating of the membranes with albumin had an inhibitory effect on oxygen radical production by all the membranes for both luminol and lucigenin-enhanced CL. Thus albumin adsorption appears to have a protective effect on cellular activation by the membranes tested. The adsorption of proteins was further analyzed by electrophoresis of the proteins removed after 10 min incubation of the membranes in fresh plasma[10]. The recovery of proteins was by far the largest on PS followed by PAN, CU, CA and PMMA (table 1 in 10) and the pattern of these proteins on electrophoresis gel (fig.1 in 10) showed that all the membranes adsorbed albumin although PS and PAN to a lesser degree. A protein of 45kD was seen on PS only, whereas a protein of 12-14 kDa was observed on PS and PAN. The results demonstrate that plasma proteins were differentially adsorbed on various dialyzer membranes and that these proteins modified the ability of the membranes to activate neutrophil oxidative metabolism.

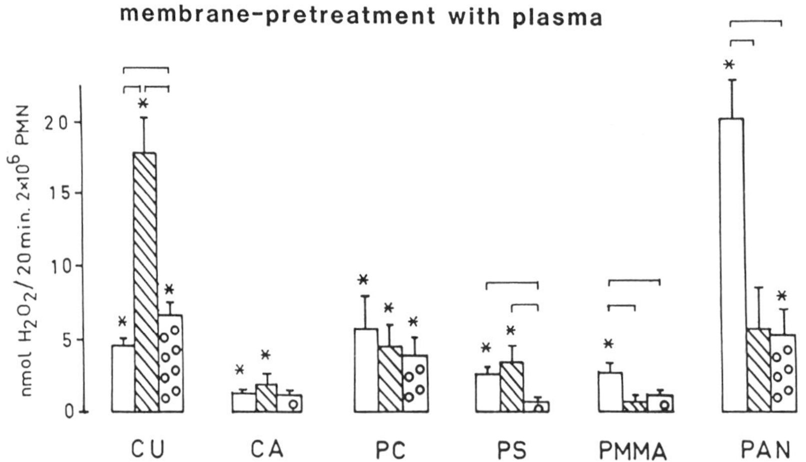

Fig.2.   In vitro hydrogen peroxide production by dialyzer membranes. Symbols are: ▢ saline-washed membranes, ▨ fresh plasma-treated membranes, ▣ heat-inactivated plasma-treated membranes. Asteriks indicate the level of significance in comparison to controls (absence of membrane) and brackets indicate the level of significance between the groups (p < 0,05).
Values are the mean ± SEM of at least 4 experiments

Table 1. Effect of membrane-albumin pretreatment on neutro-
          phil oxygen radical production

		CU	CA	PS	PAN
Luminol CL[a]	-BSA(11)	10,4±2,7	6,0±1,1	1,2±0,2	26,3±4,8*
	+BSA(6)	4,6±0,6	0,8±0,2	0,8±0,1	8,1±5,6*
Lucigenin CL[a]	-BSA(7)	2,6±0,8	7,1±1,3*	0,3±0,1	--
	+BSA(3)	0,2±1,2	0,2±0,0	0,2±0,0	--

Saline washed membranes were incubated in saline or in bovin-
serum albumin (BSA) (4%) for 10 min, washed 3 times with
saline, then assayed as described in methods.
Results are the mean ± SEM of the number of experiments in
brackets
[a]: mV
* = < 0,05 versus control (absence of membranes)

In vivo effect of membrane contact

        To further analyze the mechanism by which dialyzed neu-
trophils become functionally defective at maximum leukopenia,
leukocyte count, complement activation and oxidative metabolism were simultaneously studied at the arterial and venous
sites during dialysis with 3 membranes[6]. The data are summarized in table 2. As expected, significant changes in leukocyte count were observed on CU and PC, but no differences
were seen between arterial and venous sites. Complement activation as measured by $C_{3a\ des\ arg}$ accumulation did not differ
between the 3 membranes before dialysis (time 0 values). At 1
min, a slight increase was observed on CU and PC at the arterial site, while a further increase was seen at 1 min at the
venous site on CU and PS. At 15 min, a further increase in
complement activation was observed at the arterial site on CU
and PC, while on PS, the complement level returned back to
time zero value. At 15 min, but at the venous site, a very
high level of $C_{3a\ des\ arg}$ was found on CU and PC. On PS, it
increased to the same level to that observed after 1 min. At
maximum complement activation and leukopenia it can be seen
that neutrophils displayed a normal oxidative burst as measured by lucigenin-amplified CL after stimulation with opsonized zymosan before exposure to the CU dialyzer. On the other
membranes, no functional defect was observed. The fact that
the cells responded normally at the arterial site of the
dialyzer suggests that the neutrophils become downregulated
within the dialyzer and not in the general circulation[6].
Complement remains a candidate for this downregulation
within the dialyzer, since there was a large increase of complement activation at the outlet compared to the inlet of the
dialyzer. However, despite a large complement activation on
PC, no significant changes in neutrophil oxidative metabolism
were observed in this study.

Table 2. Complement activation and neutrophil oxidative metabolism at the arterial (A) and venous sites (V) of the dialyzer

		WBC count ($10^9/l$)	C_{3a} (ng/ml)	Lucigenin CL (mV)
CU	0	6,2±1,8	300±78	6,2±2,6
	1 min A	4,4±1,7	445±114	8,0±4,7
	1 min V	5,5±5,0	856±363	7,8±6,8
	15 min A	1,0±0,3	1148±376	4,4±2,1
	15 min V	1,2±0,4	3358±1413	1,9±1,2
PC	0	4,6±1,3	246±67	7,3±0,8
	1 min A	3,9±1,3	460±168	8,2±1,1
	1 min V	4,3±1,6	472±265	8,6±1,9
	15 min A	2,3±0,9	904±484	7,4±2,7
	15 min V	2,3±3,1	2023±1075	8,2±3,7
PS	0	6,7±1,8	240±11	9,2±3,7
	1 min A	6,2±2,5	294±48	9,2±3,1
	1 min V	6,8±2,9	553±171	9,4±3,9
	15 min A	6,1±2,6	365±49	10,5±3,2
	15 min V	6,5±3,1	609±117	10,4±3,2

Results are the mean ± SD of 3-5 experiments. Brackets indicate the level of significance at p < 0,05.

Reuse of the CU dialyzer has repeatedly been demonstrated as ameliorating both leukopenia and complement activation[5]. This improved response was shown, at least in part, to result from deposition of C_{3b} on the membrane, thus blocking complement activating sites. To investigate the role of complement in the blood membrane interaction in vivo, CU dialyzer membrane was pretreated with either a solution of human albumin, human immunoglobulins or total human plasma proteins[8]. Leukopenia was only improved with plasma proteins, while the oxidative metabolism defect of dialyzed cells seen on CU was corrected with any of the treatments. In contrast, complement activation was not prevented with any of the treatments[11]. These results show that leukocyte count and functions remain normal despite a large complement activation.

In conclusion, during dialysis with cuprophan membranes, activation of complement was not the only possible mechanism leading to leukopenia and neutrophil oxidative metabolism defect. Oxygen radical production, cellular responses, the nature of the membranes, protein adsorption might all play an early role either by themselves or in synergy. Thus the mechanism(s) leading to hemodialysis-induced leukopenia is probably the final result of indicate different pathways.

REFERENCES

1. P.R. Craddock, J. Fehr, A.P. Dalmasso, K.L.Brigham, and H.S. Jacob, Hemodialysis leucopenia: pulmonary vascular leukostasis resulting from complement activation by dialyzer cellophan membranes, <u>J. Clin. Invest.</u> 59: 879-888 (1977).

2. D.E. Chenoweth, Complement activation during hemodialysis: clinical observations, proposed mechanisms and theoretical implications, <u>Artif. Organs</u> 8: 281-287 (1984).

3. C. Heierli, M. Markert, P.H. Lambert, T. Kuwahara, and J.P. Wauters, On the mechanisms of haemodialysis-induced neutropenia: a study with five new and re-used membranes. <u>Nephrol. Dial. Transplant.</u> 3: 773-783 (1988).

4. M. Markert, C. Heierli, T. Kuwahara, J. Frei, and J.P. Wauters, Dialyzed polymorphonuclear neutrophil oxidative metabolism during dialysis: a comparative study with 5 new and reused membranes, <u>Clin. Nephrol.</u> 29: 129-136 (1988).

5. M. Markert, P.A. Waridel, C. Heierli, J.P. Wauters, Neutrophil functions during hemodialysis, <u>Contr. Nephrol.</u> 62: 99-108 (1988).

6. P. Neveceral, M. Markert, and J.P. Wauters, Neutrophil behavior during hemodialysis, <u>Trans. Am. Soc. Artif. Intern. Organs</u> XXXIV: 564-567 (1988).

7. E. Pick, and Y. Keiser, A simple colorimetric method for the measurement of hydrogen peroxide produced by cells in culture, <u>J. Immunol. Meth.</u> 38: 161-170 (1980).

8. P. Neveceral, M. Markert, J.P. Wauters, Protein adsorption on a cellulosic membrane modifies selectively leukopenia and alterations of neutrophil functions during hemodialysis, <u>Kidney Int.</u> 36:322 (1989).

9. T. Kuwahara, M. Markert, and J.P. Wauters, Neutrophil oxygen radical production by dialysis membranes, <u>Nephrol.Dial. Transplant.</u> 3: 661-665 (1988).

10. T. Kuwahara, M. Markert, and J.P. Wauters, Proteins adsorbed on hemodialysis membranes modulate neutrophil activation, <u>Artif. Organs</u> 13: 427-431 (1989).

11. P. Neveceral, M. Markert, and J.P. Wauters, manuscript <u>in preparation</u>.

BLOOD FLOW DEPENDENT GRANULOCYTE ACTIVATION

IN MEMBRANES WITH AND WITHOUT COMPLEMENT ACTIVATION

Joachim Böhler, Johannes Donauer
Peter Schollmeyer, Walter H. Hörl

Department of Medicine, Division of Nephrology
University of Freiburg, 7800 Freiburg, FRG

INTRODUCTION

Granulocyte activation during hemodialysis with cuprophane membranes has conclusively been shown to be mediated by complement activation inside the hemodialyzer[1,2]. Synthetic hemodialysis membranes cause little alteration of the complement system as indicated by low levels of C3a[3], C5a[4], or C5b-9[5] at the hemodialyzer blood outlet. However, we previously demonstrated that one of these synthetic membranes made from polymethylmethacrylate (PMMA), very strongly activates granulocytes in patients during HD-treatment despite minimal complement activation[6]. To determine whether this observation is specific for the PMMA polymer or seen in other synthetic membranes as well, we studied 3 synthetic membranes in comparison to cuprophane.

Two sets of in vitro experiments were carried out in order to elucidate the mechanism involved in granulocyte activation with or without complement activation: Different membrane materials were incubated with whole blood to compare effects of the chemical composition of polymers indepently of blood flow phenomena. The second set of experiments consisted of in vitro dialyses where blood flow through different hemodialyzers was kept constant at 200 ml/min.

METHODS

Commercially available hemodialyzers were used for both experiments. They were made from hollow fiber capillaries of 4 different polymers: cuprophane (CUP), polymethylmeth-acrylate (PMMA), polysulfone (PS), and polyacrylonitrile (PAN). Details on dialyzer modules are given in table 1.

New Aspects of Human Polymorphonuclear Leukocytes
Edited by W.H. Hörl and P.J. Schollmeyer, Plenum Press, New York, 1991

Table 1. Hemodialyzer modules used for recirculation and incubation studies (m^2 = internal surface area).

Membrane	Dialyzer m^2	Manufacturer
Cuprophane (CUP)	Hemoflow E3 1.25	Fresenius Germany
Polymethyl-(PMMA) methacrylate	Filtryzer 1.2 B2-1.2	Toray Japan
Polyacrylo- (PAN) nitrile	PAN-200 1.4	Asahi Japan
Polysulfone (PS)	F60 1.25	Fresenius Germany

1. Incubation experiments

To incubate membrane material with whole blood, hemodialyzers were first rinsed with normal saline exactly as for use in patients and then opened. Capillaries were removed from the devices and dried before weighing. 125 mg of membrane material was placed into test tubes. Venous blood was drawn from healthy donors and anticoagulated with heparin (20 U/ml). 6 ml aliquots were then added to the test tubes containing membrane material and, as control, to a test tube without dialyzer membrane.

Test tubes were incubated for 10 minutes in a water bath at 37 °C. Fig. 1 illustrates the setup for incubation experiments.

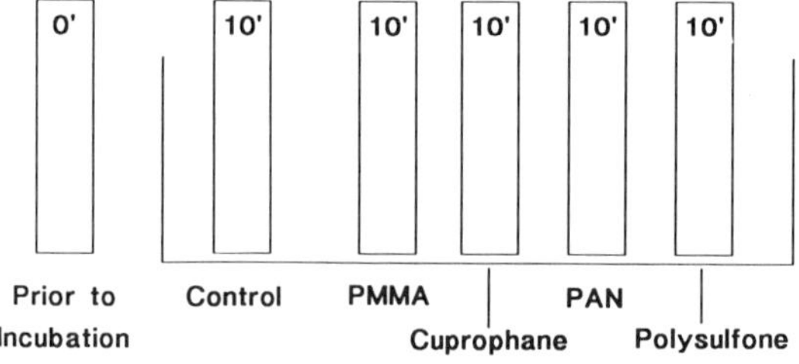

Fig. 1 Incubation of 125 mg of hemodialyzer membrane material with 6 ml heparinized blood for 10 minutes at 37 °C in a water bath. As control, 6 ml blood were incubated without membrane material.

After 10 minutes of incubation blood was withdrawn from
the test tube into an evacuated tube containing Na-EDTA for
chelation of calcium and magnesium ions in order to interrupt
ongoing activation processes. Tubes were placed on ice
immediately and centrifuged at 4 °C at 2000 x g for 10
minutes thereafter. Plasma was removed, rapidly frozen in
liquid nitrogen, and stored at -70 °C until analysis (see
below).

2. Recirculation model

To evaluate the same membranes under blood flow
conditions we used a recirculation model of hemodialysis
where blood and dialysate were pumped through commercially
available hemodialyzers (table 1). Fig. 2 illustrates the
closed loop system for dialysate and blood. After preparation
of the dialyzer exactly as for use in patients, the system
was filled with heparinized venous blood from healthy donors.
A roller pump propelled the blood toward a reservoir and from
there through the dialyzer capillaries. The blood then left
the dialyzer and entered a second reservoir. From this
'venous' reservoir the pump took the blood for the next
cycle. The dialysate loop was opperating accordingly. Filling
volume of the dialysate compartment and blood compartment was
500 and 200 ml, respectively. Dialysate composition closely
resembeled that used for bicarbonate hemodialysis in
patients. With high flux dialyzers a positive pressure on the
blood side was needed to avoid backfiltration from dialysate
to blood. This positive pressure resulted in continuous
filtration from the blood to the dialysate compartment. A
modified automatic infusion pump was continuously withdrawing
dialysate/filtrate from one of the dialysate reservoirs and

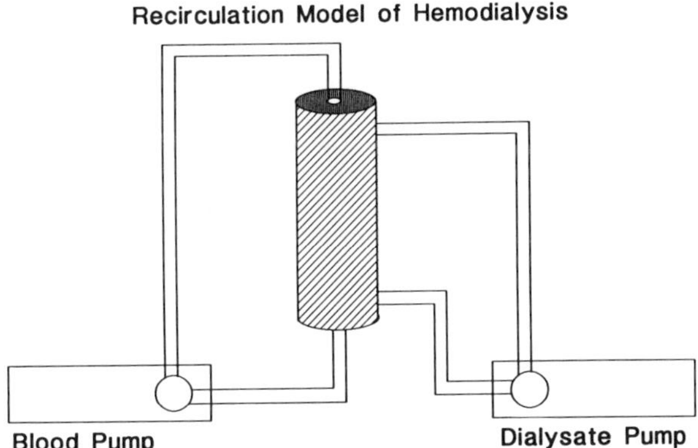

Fig. 2 Schematic representation of the closed loop
recirculation hemodialysis system. Blood and
dialysate were pumped in countercurrent fashion
by two hemodialysis roller pumps. Standard
hemodialysis tubing was used and sterilly
assembled to fit the recirculation system.

at the same rate infusing sterile substitution fluid into the blood compartment. This assured constant fluid volumes in both compartments during operation.

Blood samples for analyses were drawn at 0, 5, 15, 30, 60, 90, and 120 minutes into Na-EDTA containing evacuated tubes and processed as described above. An equal volume of substitution fluid was added to the blood compartment after each blood draw to avoid decreasing recirculation volume.

Assay procedures

Plasma levels of granulocyte elastase in complex with α_1-proteinase inhibitor (E-α_1PI) were determined by ELISA (enzyme-linked-immunosorbent-assay[7], Merck, FRG). C3a was measured by RIA (radioimmunoassay[8], Ammersham, FRG).

RESULTS

C3a during incubation

After 10 minutes of incubation of whole blood at 37°C without membrane material C3a levels (fig. 3a) had slightly increased from 296.1 (± 27.2, SEM) ng/ml (data not shown in fig. 3) to 366.4 (± 29.9) ng/ml (=control). Incubation with dialyzer membrane materials caused markedly different generation of C3a. Cuprophane increased levels of C3a 9fold compared to control (from 366.4 ± 29.9 ng/ml to 3,077 ± 243.6 ng/ml), while on the other hand presence of PAN material (613.4 ± 38.0 ng/ml) less than doubled C3a concentrations. PMMA (1,017 ± 105.9 ng/ml) and polysulfone (1,340 ± 170.7 ng/ml) caused a 3-4 fold increase of levels. In figure 3a results are shown in order of decreasing complement activating activity of the membranes (CUP >> PS > PMMA > PAN).

C3a during recirculation

C3a levels (n=4) before (=control, 388.4 ± 72.0 ng/ml) and after 90 minutes of recirculation in the closed loop in vitro dialysis system are shown in fig. 4a. Differences between membranes are even more pronounced in this model. PAN (191.0 ± 69.3 ng/ml) and PMMA (254.3 ± 50.2 ng/ml) caused no elevation of C3a levels, instead these membranes even had a tendency to lower predialysis levels of C3a. Polysulfone (1,308 ± 329.7 ng/ml) and cuprophane (13,217 ± 1,726 ng/ml) showed an increase in C3a levels. Cuprophane induced anaphylatoxin levels were about 10 times higher than with polysulfone and 34 times higher than control.

Comparison of C3a levels after incubation or recirculation

When compared to incubation results the ability of membranes to activate the complement system after recirculation followed the same order (CUP >> PS >> PMMA > PAN). However, cuprophane caused a 4 times stronger activation during recirculation (13,217 ng/ml) than during incubation (3,077 ng/ml), while polysulfone was unchanged (1,308 vs. 1,340 ng/ml). PMMA and PAN elevated C3a levels to some extent during incubation but lowered levels during recirculation. With all 4 membranes little C3a was measured in the dialysate during in vitro hemodialysis (data not shown) indicating that this anaphylatoxin is dialysable, but decrease of levels with PAN and PMMA below control is most likely mainly due to adsorption to the membrane[4,9,10] rather than dialysis.

Elastase release during incubation and in vitro dialysis

Fig. 3b and 4b show measurements of elastase-α_1PI-complex in both models: cuprophane was by far the most active membrane in inducing release of elastase from granulocytes in both models, causing an incresae in E-α_1PI-complex from 42.4 ± 2.2 ng/ml (=control) to 541.1 ± 49.6 ng/ml in incubation and from 97.6 ± 8.4 ng/ml (=control) to 595.0 ± 236.8 ng/ml in recirculation. When comparing synthetic membranes, however, clear differences in the order of elastase releasing capacity were seen between models. In incubation: CUP (541.1 ± 49.6 ng/ml) >> PAN (279.4 ± 34.9 ng/ml) > PMMA (160.4 ± 19.8 ng/ml) > PS (150.6 ± 20.0 ng/ml); in recirculation: CUP (595.0 ± 236.8 ng/ml) > PMMA (549.6 ± 177.5 ng/ml) >> PS (291.5 ± 87.8 ng/ml) > PAN (176.5 ± 42.3 ng/ml). PAN ranked second to cuprophane during incubation, but was least active during recirculation. PMMA and polysulfone induced only little elastase release during incubation, but during recirculation PMMA was almost as potent in activating granulocytes as cuprophane.

Correlation between C3a and elastase levels

Studying the correlation between C3a generation and granulocyte elastase release due to membrane contact, only cuprophane consistently activated both, the complement system and granulocytes. In synthetic membranes, however, we found no correlation between complement and elastase levels. PAN, causing little complement activation in both models induced significant elastase release during incubation but not during recirculation. In contrast, PMMA prompted a very pronounced rise in elastase levels during recirculation, despite its capacity to even decrease C3a concentrations below control levels.

DISCUSSION

Based on early studies by Craddock[1] and Hammerschmidt[2] in cuprophane membranes complement activation is believed to be the main mechanism by which granulocytes are activated during hemodialysis. Newer synthetic membranes (PS, PMMA, PAN and others) have repeatedly been shown to induce less complement activation[3,11] and less granulocytopenia[3,12].

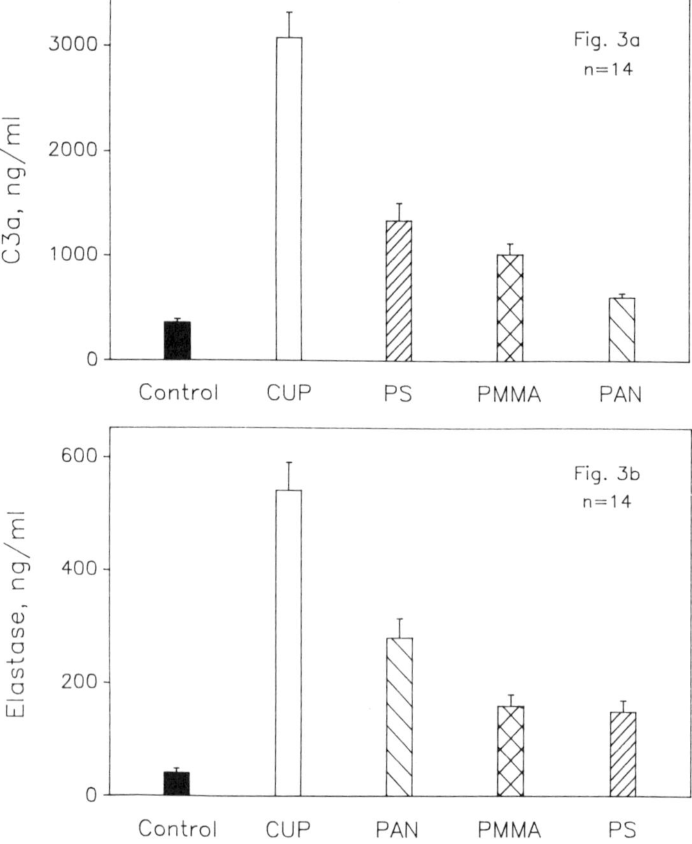

Fig. 3 Levels of C3a (upper pannel) and of elastase in complex with α_1-proteinase inhibitor (E-α_1PI, lower pannel) after incubation of blood with membrane material for 10 minutes at 37 $^{\circ}$C. Blood incubated without membrane material served as control.

With use of more sophisticated parameters of granulocyte activation, it has been demonstrated that, although to a lesser extent than in cuprophane, synthetic membranes may also activate leukocytes[13]. One of the mechanisms held responsible, at least for monocyte activation, is backdiffusion of unsterile dialysate into the blood compartment during hemodialysis in patients[14]. This possible cause of activation was excluded in our study because sterile and pyrogen free dialysate was used. Nevertheless, depending on the membrane type and investigational model marked activation of granulocytes did occur in synthetic membranes.

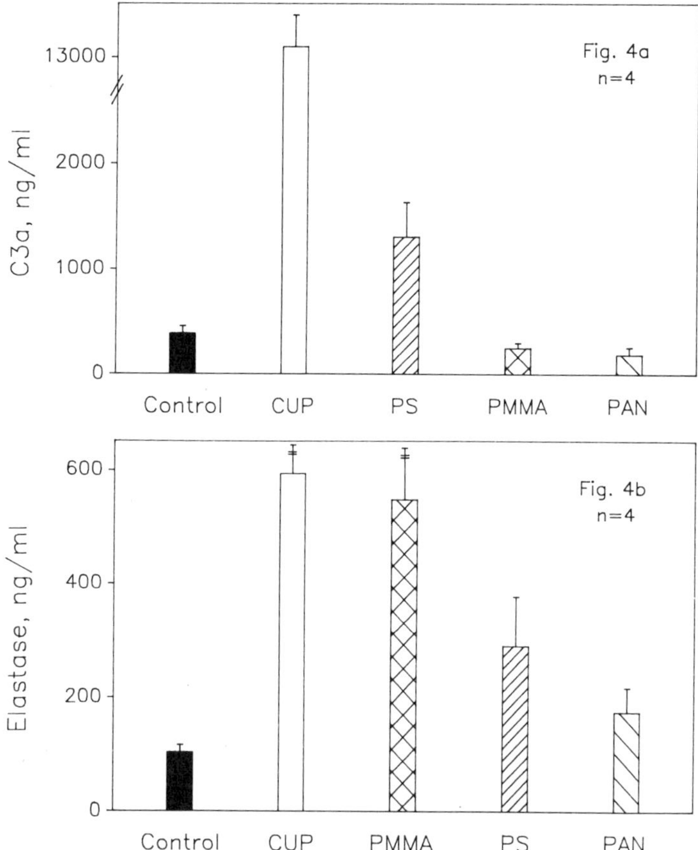

Fig. 4 Levels of C3a (upper pannel) and of elastase in complex with α_1-proteinase inhibitor (Eα_1PI, lower pannel) after 90 minutes of recirculation of heparinized blood through four capillary hemodialyzers made from different polymeric membranes. As control a blood sample not in contact with membrane material was used.

Comparison of data from patient studies, in vitro dialysis, and incubation experiments has to observe the intrinsic differences between these models. In patients, modification of substances and cells leaving the dialyzer profoundly influences measured parameters. In hemodialysis with cuprophane membranes most of the activated granulocytes that embolize to the lungs[15] have never passed through the dialyzer, instead they are activated by mediators (C3a and C5a) generated inside the device and released into the circulation. These phenomena obviously can not occur in in vitro models. Overlay of true membrane effects with reactions of the body can thus be prevented.

Incubating blood with capillaries removed from hemodialyzers, brings plasma and cells in contact with the outer surface of the capillaries, while normally blood is directed over the smooth inner surface of the capillary. The chemical composition of the capillaries is identical on both sides, but mechanical properties will differ and shear stress has to be avoided in incubation. Capillaries were compared on a per weight basis (125 mg per 6 ml blood), but surface area may differ, since capillaries of different polymers are not of the same diameter.

Polysulfone material induces some complement activation in patients[11], with incubation (fig. 3a), and during in vitro dialysis (fig. 4a). In parallel some elastase release from granulocytes is observed. With C3a and elastase detected simultaneously in this membrane a complement mediated activation of granulocytes is one of the possible explanations. Complement independent mechanisms, however, can not be excluded on the basis of our data.

PAN induced elastase release during incubation but almost no granulocyte activation during recirculation. This may reflect the larger capacity to bind C3a and elastase in the recirculation system. PAN known to avidly adsorb various plasma proteins[4,9], caused C3a levels higher than control in incubation but lower than baseline after recirculation. The blood to membrane ratio (by weight) was higher in incubation, which possibly accounted for more generation than adsorption of activation products. During hemodialysis in patients, effects similar to in vitro dialysis have been reported[9,10].

Our earlier results in patients[6] and the in vitro studies presented here demonstrate that PMMA membrane induces granulocyte activation in the absence of elevated anaphylatoxin levels. Low levels of complement fragments in the blood leaving the dialyzer may indicate absence of complement activation inside the dialyzer or activation of the complement cascade but concurrent adsorption to the membrane or removal of complement fragments by dialysis. In this letter instance, complement mediated granulocyte activation may be present in the absence of elevated levels of C3a. Since C3a levels even decreased during in vitro dialysis with PMMA, this membrane is obviously capable of clearing C3a from the circulation, a phenomenon well known for PAN[4,9]. Despite adsorption of proteins, it is unlikely that PMMA will produce a complement dependent activation of granulocytes almost as strong as cuprophane, with no complement fragments detectable. Instead, in our opinion PMMA is an example of a synthetic membrane which potently activates granulocytes by a complement independent mechanism. Our data also suggest that this mechanism may depend on blood flow, since during incubation only limited elastase release was seen, while during recirculation at a blood flow of 200 ml/min PMMA induced elastase relase was almost as high as in cuprophane membranes.

SUMMARY

Synthetic membranes activate the complement system to some degree but also clear complement fragments from the circulation by adsorption and/or filtration and dialysis, PAN and PMMA even lower preexisting plasma levels of C3a. Activation of granulocytes measured as elastase release does occur with all membranes and might be due to complement activation. However, in synthetic membranes no correlation exists between levels of complement fragment C3a and elastase release. Instead, even when PMMA membrane lowered preexisting plasma levels of C3a below baseline, activation of granulocytes to a similar degree as known from cuprophane membranes was observed. This activation did occur whith blood flow of 200 ml/min during recirculation, but not without blood flow during incubation of membrane material.

We conclude that granulocyte activation in synthetic membranes does not require an activated complement system, but may be a blood flow dependent phenomenon.

REFERENCES

1. Craddock PR, Fehr J, Dalmasso AP, Brigham KL, Jacob HS. Hemodialysis leukopenia. Pulmonary vascular leukostasis resulting from complement activation by dialyzer cellophane membranes. <u>J Clin Invest</u> 1977; 59: 879-888.
2. Hammerschmidt DE, Harris PD, Wayland JH, Craddock PR, Jacob HS. Complement-induced granulocyte aggregation in vivo. <u>Amer J Pathol</u> 1981; 102: 146-150.
3. Chenoweth DE, Cheung AK, Ward DM, Henderson LW. Anaphylatoxin formation during hemodialysis: effect of different dialyzer membranes. <u>Kidney Int</u> 1983; 24: 764-769.
4. Jorstad S, Smeby LC, Balstad T, Wideroe TE. Generation and removal of anaphylatoxins during hemofiltration with five different membranes. <u>Blood Purif</u> 1988; 6:325-335.
5. Deppisch R, Schmitt V, Bommer J, Hänsch GM, Ritz E, Rauterberg EW. Fluid phase generation of terminal complement complex as a novel index of bioincompatibility. <u>Kidney Int</u> 1990; 37: 696-706.
6. Hörl WH, Riegel W, Schollmeyer P, Rauterberg W, Neumann S. Different complement and granulocyte activation in patients dialyzed with PMMA dialyzers. <u>Clin Nephrol</u> 1986; 25: 304-307.
7. Neumann S, Hennrich N, Gunzer G, Lang H. Enzyme-linked immunoassay for human granulocyte elastase in complex with α_1-proteinase inhibitor, <u>in</u> Proteases: Potential Role in Health and Disease, Hörl WH, Heidland A (eds), Plenum Press, New York, London 1984, pp 379-390.
8. Hugli TE, Chenoweth DE. Biologically active peptides of complement: Techniques and significance of C3a and C5a measurements. <u>in</u>: Immunoassays: Clinical Laboratory Techniques for the 1980s, Nakamura RM, Dito WR, Tucker III, ES (eds), Alan R Liss Inc. New York 1980, pp 443-460.

9.   Kandus A, Kladnik S, Ponikvar R, Ivanovich P, Drinovec J. Adsorption of anaphylatoxins C3a and C5a on AN69 membrane of hollow fiber and plate dialyzer - in vivo study (Abstract). <u>Blood Purif</u> 1988; 6:355.

10.   Röckel A, Hertel J, Perschel WT, Walb D, Fiegel P, Panitz N, Abdelhamid S. Polyacrylonitrile complement activation attenuated by simultaneous ultrafiltration and adsorption. <u>Nephrol Dial Transplant</u> 1987; 2: 251-253.

11.   Klinkmann H, Falkenhagen D. Biocompatibility - a system approach. <u>in</u>: New Perspectives in Hemodialysis, Peritoneal Dialysis, Arteriovenous Hemofiltration, and Plasmapheresis, Hörl WH, Schollmeyer PJ (eds), Plenum Press, New York, London 1989, pp 39-52.

12.   Hakim RM, Fearon DT, Lazarus JM. Biocompatibility of dialysis membranes: effects of chronic complement activation. <u>Kidney Int</u> 1984; 26:194-200.

13.   Hörl WH, Steinhauer HB, Schollmeyer P. Plasma levels of granulocyte elastase during hemodialysis: Effects of different dialyzer membranes. <u>Kidney Int</u> 1985; 28:791-796.

14.   Lonnemann G, Bingel M, Floege J, Koch KM, Shaldon S, Dinarello CA. Detection of endotoxin-like interleukin-1-inducing activity during in vitro dialysis. <u>Kidney Int</u> 1988; 33: 29-35.

15.   Brubaker LH, Jensen D, Johnson C, Nothum R, Nolph KD. Kinetics of stagnation-induced neutropenia during hemodialysis. <u>Trans Amer Soc Artif Intern Organs</u> 1972; 18: 305-311.